ANATOMIC GUIDE FOR THE ELECTROMYOGRAPHER

ANATOMIC GUIDE FOR THE ELECTROMYOGRAPHER

THE LIMBS

by

EDWARD F. DELAGI, M.D.
JOHN IAZZETTI, M.D.
ALDO PEROTTO, M.D.
DANIEL MORRISON, M.D.

Second Edition, Second Printing by

EDWARD F. DELAGI, M.D.
Professor of Rehabilitation Medicine
Albert Einstein College of Medicine
New York, New York

ALDO PEROTTO, M.D.
Associate Professor of Rehabilitation Medicine
Albert Einstein College of Medicine
New York, New York

With a Foreword by

Arthur S. Abramson, M.D.
Professor and Chairman, Department of Rehabilitation Medicine
Albert Einstein College of Medicine of Yeshiva University
New York, New York

Illustrated by **Hugh Thomas**

CHARLES C THOMAS • PUBLISHER
Springfield • Illinois • U.S.A.

Published and Distributed Throughout the World by

CHARLES C THOMAS • PUBLISHER
2600 South First Street
Springfield, Illinois, 62717, U.S.A.

ISBN 0-398-03951-8

Library of Congress Catalog Card Number: 79-15906

First Edition, 1975
Second Edition, First Printing, 1980
Second Edition, Second Printing, 1981

With THOMAS BOOKS *careful attention is given to all details of manufacturing and design. It is the Publisher's desire to present books that are satisfactory as to their physical qualities and artistic possibilities and appropriate for their particular use.* THOMAS BOOKS *will be true to those laws of quality that assure a good name and good will.*

Printed in the United States of America
R-1

Library of Congress Cataloging in Publication Data

Delagi, Edward F
Anatomic guide for the electromyographer—the limbs.

First ed. by E. F. Delagi and others published in 1974.
Bibliography: p.
1. Electromyography. 2. Extremities (Anatomy)
I. Perotto, Aldo, joint author. II. Anatomic guide for the electromyographer: the limbs. III. Title.
[DNLM: 1. Electromyography. 2. Extremities—Anatomy and histology. WE500. 3 D333a]
RC77.5.A5 1979 616.7'4'0754 79-15906
ISBN 0-398-03951-8

FOREWORD

Electromyography has been developing as an increasingly useful diagnostic tool for only the past thirty years. An abundance of literature has given the clinician an opportunity to become more and more sophisticated in its use. Unfortunately, understanding the tool as a diagnostic aid rather than as a diagnosis maker has not always been evident. It is the knowledgeable and thorough clinician who is best fitted to realize the great power of the procedure. On the other hand, having a detailed knowledge of electromyography by itself does not necessarily make one a good clinician. The reason for this is quite simple. *Whether, when* and *where* the procedure should be done are based on knowledge, experience and judgment—these are the hallmarks of the clinician. If done, the findings must be interpreted in light of the patient's problem as previously elicited from history, physical examination and other diagnostic tests.

This book is largely concerned with the question of *where* to insert electrodes in order to elicit maximum useful information. The more experienced the electromyographer becomes, the more he realizes the need for precision in the placement of needle electrodes. After all, there are very many anatomically distinct skeletal muscles in the human body. Each of them, whether large or small, superficial or deep, is innervated by axones traversing roots, plexuses and nerve trunks. In addition, innervations are often abberant. One example of the diagnostic power of this procedure is the ability to localize discrete axonal lesions within reasonably narrow zones. This depends upon eliciting electromyographic abnormalities in distinctive combinations of a number of individual muscles. This example illustrates the need for the clinician to be a good anatomist. Unfortunately this is not always the hallmark of even the exceptional clinician, and the modern deemphasis of detailed anatomical study in our medical schools is of no help. This poses no problem since the authors have done the job for us. This simple book, almost in the nature of a cookbook, is the product of painstaking detailed and repeated dissections by the authors. It is a labor of love–but who could have carved out the long hours from busy professional and academic lives except those in love with their field.

Segments of this work had been presented at seminars in national meetings. These presentations had been received with enthusiasm and with appeals for a more complete and tangible presentation of the material. For the clinician who wishes to make better syntheses of his patients' problems, this book is offered as an aid to a diagnostic aid.

Arthur S. Abramson, M.D.

PREFACE TO SECOND EDITION

In revising this manual we have tried to take into consideration the many constructive criticisms offered by our readers. These suggestions fell into three main categories:

The first was for a clearer exposition of the innervation of the muscles presented. We attempt to do this by giving the course of each innervation from the periphery through the peripheral nerves, cords, division, trunks and spinal segments. Where more than one spinal segment contributes to the innervation, we have italicized the segment or segments making the major contribution.

Another suggestion we thought worthy of attention was to have cross-sectional diagrams of some of the deep muscles in order to aid in the orientation of the reader in placing the electrode. We have added such diagram for the pronator quadratus, the flexor pollicis longus, the extensor digitorum longus, the tibialis posterior, the adductor brevis and the vastus intermedius.

Finally, we were asked to be more specific in listing our literary sources than stated in the first edition. To fulfill this request we add the following sources:

(1) Gross, Charles M. (Ed.): *Gray's Anatomy of the Human Body,* 29th American ed. Lea & Febiger, Philadelphia,, 1973.

(2) Baker-Cohen, Frances: *A Study Guide and Laboratory Manual for Human Anatomy,* corrected and expanded edition, Albert Einstein College of Medicine, New York, 1971.

(3) Haymaker, Webb, and Woodhall, Barnes: *Peripheral Nerve Injuries,* 2nd ed. Saunders, Philadelphia, 1953.

(4) Hollinshead, Henry: *Functional Anatomy of Limbs and Back,* 2nd ed. Saunders, Philadelphia, 1960.

(5) Sunderland, Sydney: *Nerve and Nerve Injuries.* Williams & Wilkins, Baltimore, 1968.

(6) Crenshaw, A.H. (Ed.): *Campbell's Operative Orthopaedics,* 4th ed. Mosby, St. Louis, 1963.

(7) Spinner, Morton: *Injuries to the Major Branches of Peripheral Nerves of the Forearm.* Saunders, Philadelphia, 1972.

(8) Lockhart, R.D., Hamilton, G.F., and Fyfe, F.W.: *Anatomy of the Human Body.* Lippincott, Philadelphia, 1965.

(9) Kakamo, Kenneth: The Entrapment Neuropathies. *Muscle and Nerves, 1:*264-279, July-Aug 1978.

(10) Lampe, Ernest W.L.: Surgical anatomy of the hand. *Clinical Symposia–Ciba, 21(3):* July-Aug-Sept 1969.

(11) Koppel, Harvey and Thompson, Walder A.L.: *Peripheral Entrapment Neuropathies.* Krieger, Huntington, 1976.

E.F.D.
A.O.P.

INTRODUCTION

Development in electronic instrumentation has made it possible for electrophysiological observations, formerly only possible in the laboratory, to become commonplace clinical diagnostic procedures. With this advance in technology, a new demand is put upon the diagnostician, not only to become acquainted with the technical aspects of the new instrumentation but to reorganize the more traditional areas of his knowledge so that they better serve this new approach.

In order to perform the now common electrodiagnostic procedures of electromyography and nerve conduction studies, a knowledge of anatomy of the muscle, nerve and skeletal systems is essential.

This book represents an attempt to supply the electrodiagnostician with anatomical information concerning the limbs, organized for easy utilization in the clinical practice of his discipline.

In normal subjects it is usually possible to use kinesiologic maneuvers to identify the muscle and localize the electrode. However, in certain movements where synergistic muscles cannot be activated discreetly, this is not possible. It is also not possible to use this method of localization in patients with complete paralysis or with neurological conditions that make it impossible to perform controlled or discreet movements.

Other methods previously described include using the electrode as a means of stimulating the muscle, radiological localization with or without the injection of carbon dioxide through the electrode, or using the motor points as the area for electrode insertion.

Although these methods are quite suitable in certain circumstances, they require special apparatus and are solely methods of localization not answering the problems of initial placement of the electrode.

The muscles accessible to the examiner have been placed in regional categories in order to facilitate clinical study. A brief anatomical description and a diagram is given for each muscle and the special requirements for the procedure as follows:

1. Segmental and peripheral innervation and origin and insertion.
2. Position of patient or limb.
3. Point of electrode insertion determined by visual and/or palpable landmarks.
4. Test maneuver to insure proper localization of electrodes after insertion.
5. Pitfalls or possible sources of error.

The above is based on a systematic anatomical study of cadaver material, normal subjects, patients and standard anatomy texts and manuals.

Although this manual is based on data derived from the anatomy of adult subjects of the usual body configuration found in patients and anatomical specimens in east coast urban centers, it is possible by interpolation to accomodate for extremes of anatomical variation in the adult and the, as yet, incomplete development of the child or infant.

There is rarely a problem when the point of insertion is given as a proportion between two boney landmarks. However, when measurements are given in terms of "fingerbreadths" and there is great disparity between the stature of the subject and examiner, it has been suggested that the fingerbreadths used should be those of the subject and not of the examiner.

ACKNOWLEDGMENTS TO FIRST EDITION

This book was made possible by many people; unfortunately only a few can be named.

Our department chairman, Doctor Arthur Abramson was most helpful, not only in guiding and encouraging us, but in finding the necessary material support for production of the manuscript.

Doctor Frances Baker-Cohen, of the department of anatomy, gave freely of her vast knowledge of the subject and made available the essential anatomical material.

Ms. Sylvia Golod cheerfully worked through the many revisions of the manuscript made necessary by the editorial ineptitudes of the authors.

Among the nameless are the numerous medical students whose dissections were used in this study and to whom we owe many apologies for disturbing their studies to foster our own. And finally, the nameless host that made the ultimate contribution; that of their earthly remains to become the "anatomical material."

To all those named and the nameless, we give profound thanks and acknowledge our debt.

The Authors

CONTENTS

Section III
ARM

Section IV
SHOULDER JOINT

Section V
SHOULDER GIRDLE

Section VI
FOOT

Section VII
LEG

Section VIII
THIGH

Section IX
PELVIS AND HIP JOINT

ANATOMIC GUIDE FOR THE ELECTROMYOGRAPHER

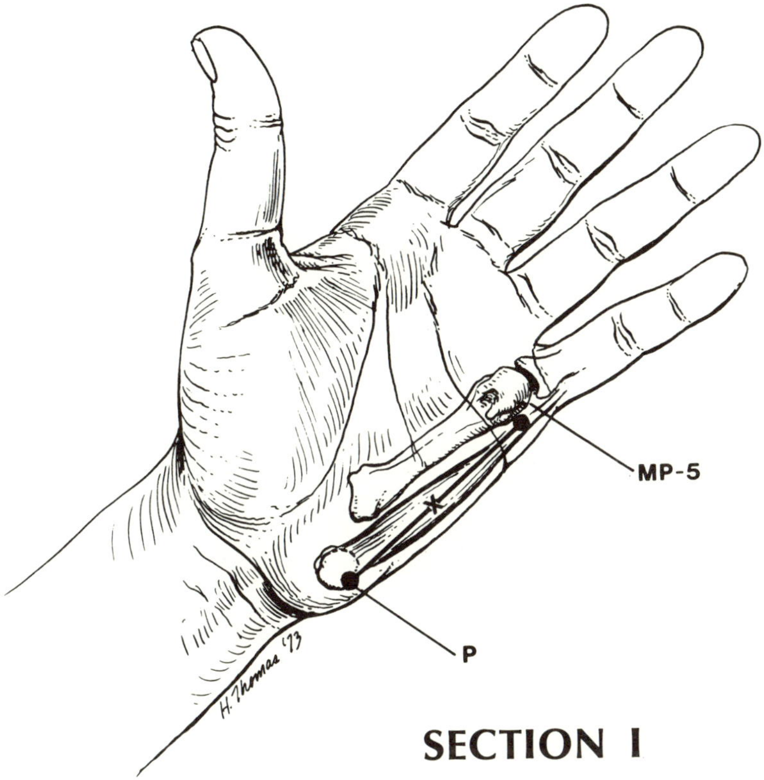

SECTION I

HAND

ABDUCTOR DIGITI MINIMI

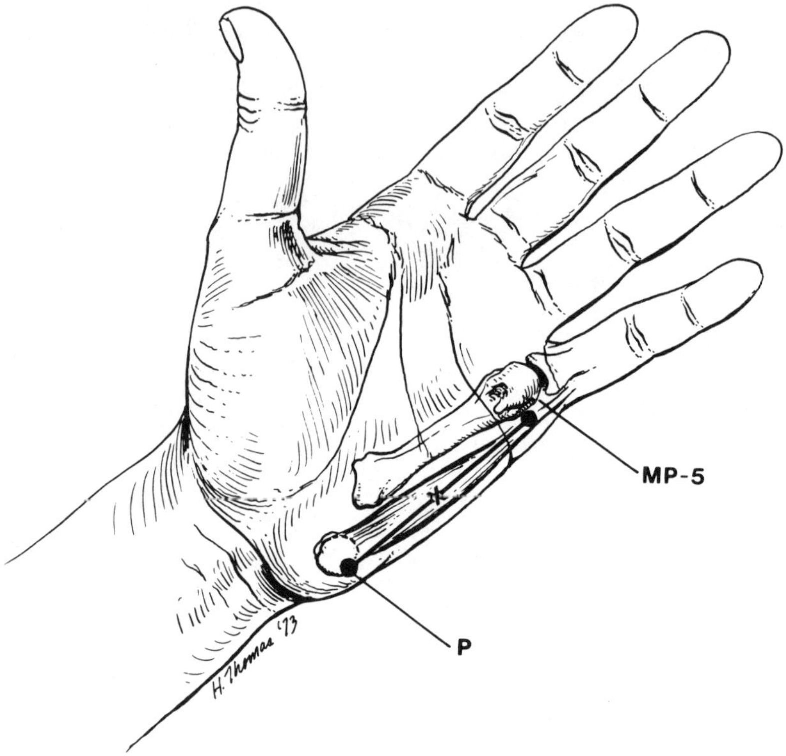

Innervation

Ulnar Nerve, Medial Cord, Anterior Division, Lower Trunk, C8, TI.

Origin

From the pisiform.

Insertion

On the medial side of the little finger into the base of the proximal phalanx.

Position

Hand in full supination.

Electrode Insertion (X)

Insert electrode to a depth of one-fourth to one-half inch at the midpoint of a line drawn between the ulnar aspects of the fifth metacarpophalangeal joint (MP-5) and the ulnar aspect of the pisiform (P).

Pitfalls

If the electrode is inserted too deeply it will be in the opponens digiti minimi.

Comment

(a) Commonly used as recording muscle for ulnar nerve motor conduction study.

(b) Involved in most ulnar nerve lesions except in Guyon tunnel entrapment when innervated through superficial palmar branch.

(c) Involved in Klumpke's palsy (avulsion of C8, TI roots).

ABDUCTOR POLLICIS BREVIS

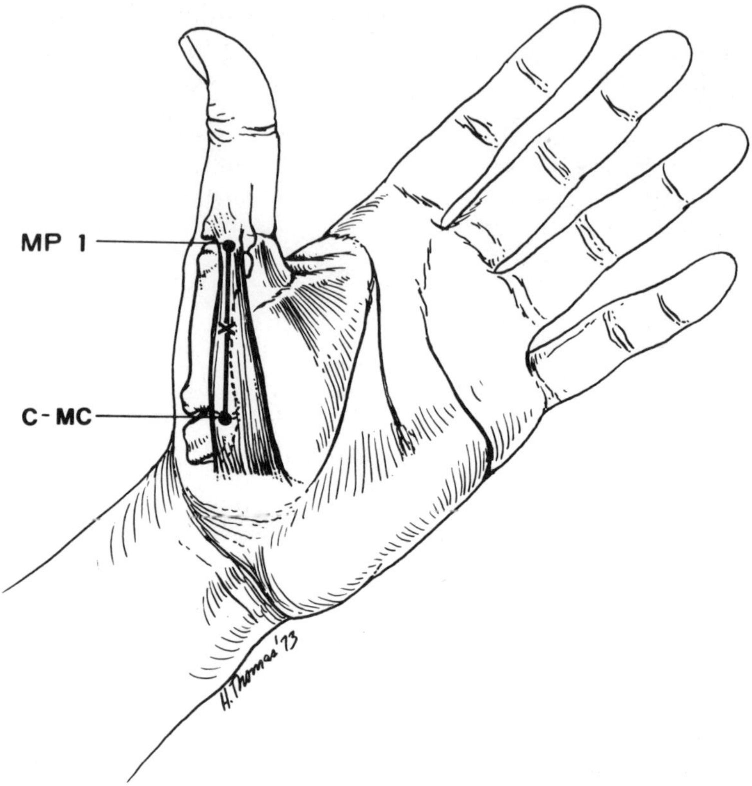

Innervation

Median Nerve, Medial Cord, Anterior Division, Lower Trunk, *C8,* TI.

Origin

From the palmar retinaculum, the tubercle of the scaphoid and that of the trapezium.

Insertion

Lateral side of the base of the proximal phalanx of the thumb.

Position

Hand in full supination.

Electrode Insertion (X)

Midpoint of a line drawn between the volar aspect of the first metacarpophalangeal joint (MP-1) and the carpometacarpal joint (C-MC). Insert to depth of one-fourth to one-half inch.

Test Maneuver

Palmar abduction of the thumb.

Pitfalls

If the electrode is inserted too deeply it will be in the opponens pollicis.

Comment

(a) Frequently used as recording muscle for median nerve motor conduction study.
(b) May be involved in all median nerve entrapment syndromes (carpal tunnel; pronator teres; ligament of Struthers) except anterior interosseus syndrome.
(c) Involved in Klumpke's palsy (avulsion of C8, TI roots).

ADDUCTOR POLLICIS

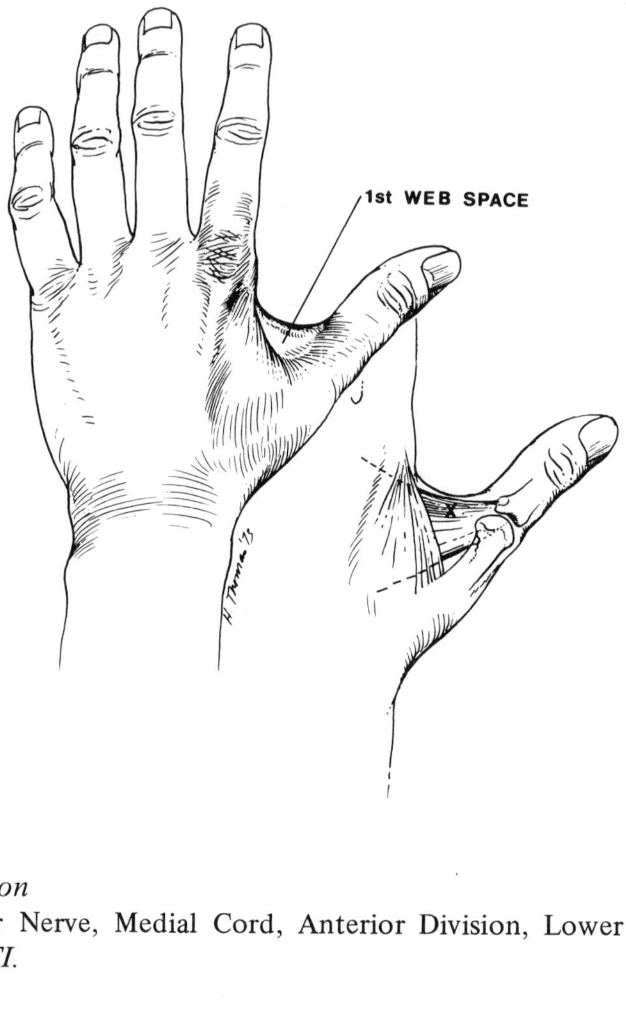

Innervation

Ulnar Nerve, Medial Cord, Anterior Division, Lower Trunk, C8, *TI.*

Origin

Lateral border of the third metacarpal.

Insertion

Medial side of the base of the proximal phalanx.

Position

Hand in full pronation, thumb in radial abduction.

Electrode Insertion (X)

At the free edge of the first web space. The neddle is directed toward the proximal end of the first metacarpal bone.

Test Maneuver

Adduct the thumb.

Pitfalls

If the electrode is inserted too dorsally it will be in the first dorsal interosseus; if too volarly it will be in the opponens pollicis.

Comment

(a) The most distal muscle innervated by the ulnar nerve.
(b) Paresis or paralysis of this muscle results in Froment's sign (substitution of flexor pollicis longus on attempted adduction of thumb).
(c) May be involved in ulnar entrapment syndromes (Guyon's Tunnel; cubital tunnel; tardy ulnar palsy; cervical rib) and Klumpke's palsy (avulsion of C8, TI nerve roots).

DORSAL INTEROSSEI

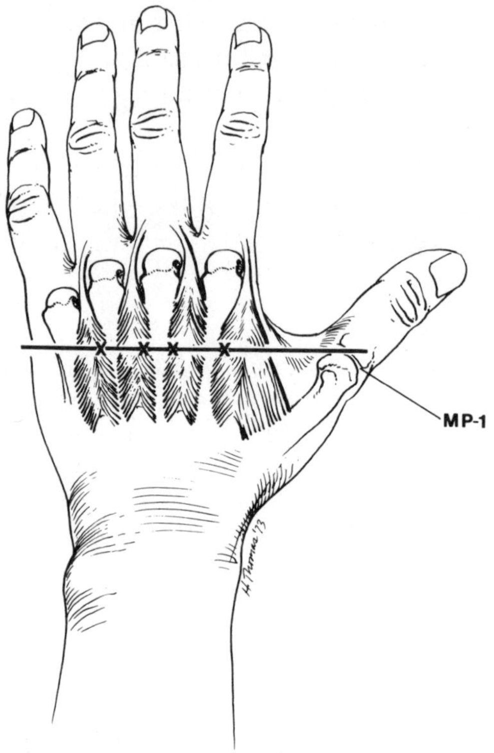

Innervation

Ulnar Nerve, Medial Cord, Anterior Division, Lower Trunk, C8, *TI.*

Origin

The first and second dorsal interossei originate on the radial aspect of the second and third metacarpal. The third and fourth dorsal interossei originate on the ulnar aspect of the third and fourth metacarpal. A small portion of the muscle originates on the opposite metacarpal.

Insertion

Base of the proximal phalanges and the dorsal digital expansions.

Electrode Insertions (X)

The common landmark for both volar and dorsal interossei is a transmetacarpal line perpendicular to the long axis of the hand at the level of the first metacarpal joint (MP-I). Insertions (Xs) are made along this line for specific interosseus muscles as indicated below:

First dorsal: Just radial to second metacarpal
Second dorsal: Just radial to third metacarpal
Third dorsal: Just ulnar to third metacarpal
Fourth dorsal: Just ulnar to fourth metacarpal

Test Maneuver

The first and second dorsal interosseus radially deviate the second and third digit, respectively. The third and fourth dorsal interosseus ulnarly deviate the third and fourth digit, respectively.

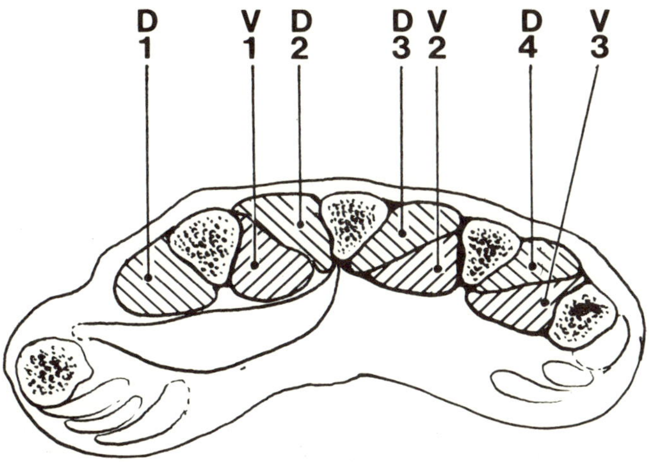

Pitfalls

First dorsal (DI): If the electrode is inserted too deeply it will be in the adductor pollicis.

Second dorsal (D2): If the electrode is inserted too deeply and it is angled in a radial direction it will be in the first volar interosseus; if deeper it will pierce the aponeurosis, and it will be in the adductor pollicis.

Third dorsal (D3): If the electrode is inserted too deeply and angled ulnarly it will be in the second volar interosseus.

Fourth dorsal (D4): If the electrode is inserted too deeply it will be in the third volar and if deeper, through the palmar aponeurosis in the opponens digiti minimi.

It is a common misconception that the dorsal (D) and volar (V) interosseus muscles lie in parallel planes, one completely over the other in the interosseous space, when in fact the interosseous space is divided obliquely with the greatest portion of the bulk of each muscle lying alongside the metacarpal of the digit upon which it acts; thus, orientation of the electrode along the radial-ulnar line becomes the critical factor rather than the dorsal-volar placement. (See figures page 11).

Comment

(a) There is a great variability innervation of these muscles so that they are sometimes innervated by either or both of the ulnar or median nerves.

(b) The first dorsal interosseus occasionally receives innervation from the musculocutaneous nerve (Sunderland).

(c) The first dorsal interosseus is used as the recording muscle in motor conduction studies of the deep palmar branch of the ulnar nerve. This is frequently involved in Guyon tunnel entrapment and is manifested by an increased latency of more than one ms. over that of the abductor digiti minimi on stimulation at the wrist.

(d) Frequently involved in Tardy ulnar palsy and Klumpke's palsy (avulsion of C8, TI nerve roots).

VOLAR INTEROSSEI

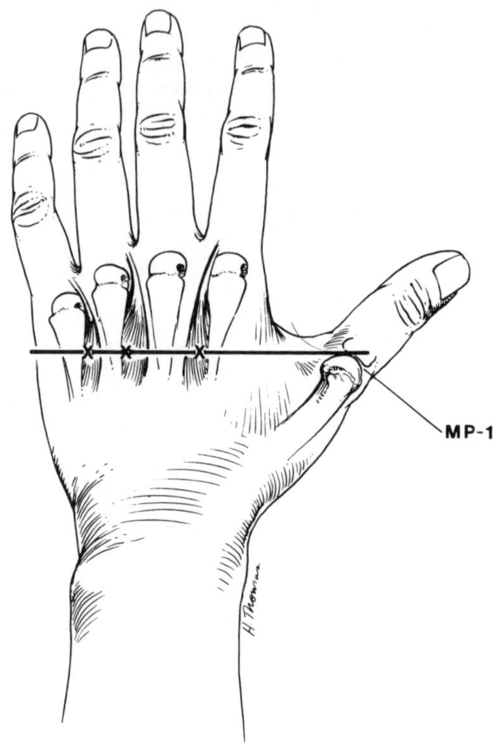

Innervation

Ulnar Nerve, Medial, Cord, Anterior Division, Lower Trunk, C8, *TI.*

Origin

The first volar arises from the ulnar aspect of the second metacarpal; the second and third volar arise from the radial aspect of the fourth and fifth metacarpal. A small portion of the muscle originates on the opposite metacarpal.

Insertion

Bases of the proximal phalanges and the dorsal digital expansions.

Electrode Insertion (X)

Along the transmetacarpal line, insert electrodes to depth of one-fourth inch.

First volar: Just ulnar to second metacarpal.

Second volar: Just radial to fourth metacarpal.

Third volar: Just radial to the fifth metacarpal.

Test Maneuver

First volar: Ulnarly deviate the second digit.

Second volar: Radially deviate the fourth digit.

Third volar: Radially deviate the fifth digit.

Pitfalls

First volar: If electrode is inserted too superficially it will be in the second dorsal interosseus; if inserted too deeply, it will be in adductor pollicis.

Second volar: If electrode is inserted too superficially it will be in the third dorsal interosseus.

Third volar: If electrode is inserted too superficially it will be in the fourth dorsal interosseus; if inserted too deeply it will be in opponens digiti minimi.

Comment

(a) These muscles show the same variability in innervation as the dorsal interossei.

(b) Frequently involved in ulnar nerve entrapment (Guyon's tunnel; cubital canal; Tardy palsy) and Klumpke's palsy (avulsion of C8, TI roots).

LUMBRICALS

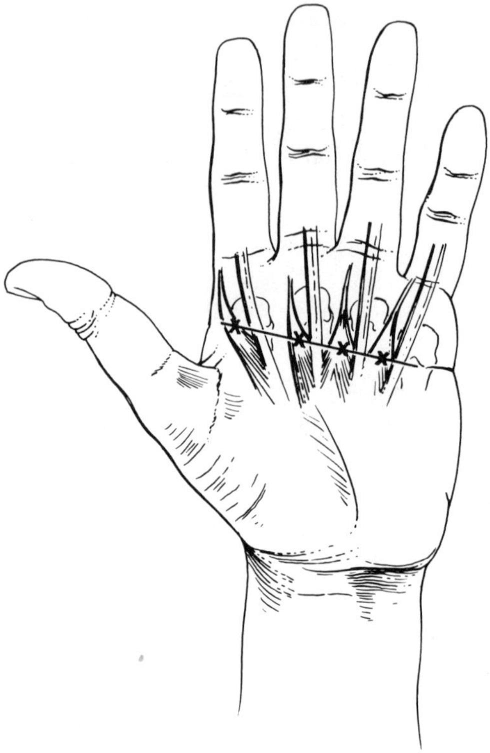

Innervation

First and Second: Median Nerve; Medial Cord, Anterior Division, Lower Trunk, *C8,* TI.

Third and Fourth: Ulnar Nerve, Medial Cord, Anterior Division, Lower Trunk, C8, *TI.*

Origin

From the radial aspect of the tendon sheath of the flexor digitoraum profundus.

Insertion

Into the radial lateral band of the dorsal digital expansion.

Position

Hand in full supination.

Electrode Insertion (X)

Just proximal to the joint and radial to the flexor tendon.

Test Maneuver

The usual method of testing lumbrical function by extending interphalangeal joints with the metacarpophalangeal joint in flexion is not possible because it physically interferes with the electrode. The preferred method is to maintain the metacarpophalangeal joints in extension and extend the interphalangeal joints against resistance.

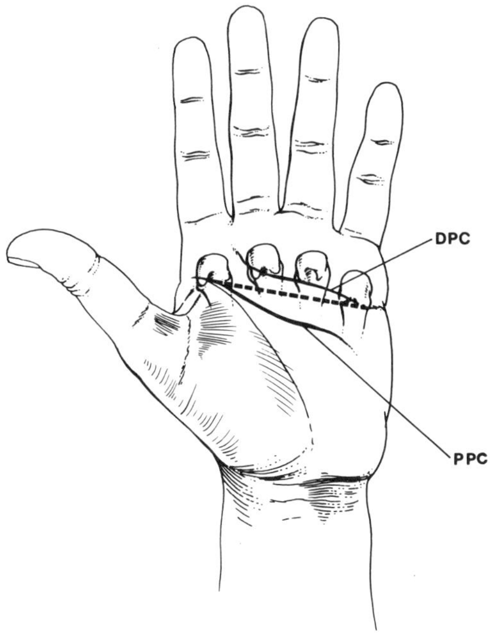

The metacarpophalangeal joints have direct relationship with both the proximal (PPC) and the distal (DPC) palmar creases. The distal crease lies over the third, fourth and fifth metacarpophalangeal joints, while the proximal lies over the second metacarpophalangeal joint. The numerical designation of a lumbrical muscle is one less than the number of the metacarpophalangeal joint it crosses, i.e. first lumbrical crosses second metacarpophalangeal joint.

Pitfalls

First Lumbrical: If the electrode is inserted too deeply it will be in the adductor pollicis; if deeper it will be in the first dorsal interosseus.

Second Lumbrical: If the electrode is inserted too deeply it will be in the most ulnar fibers of the adductor pollicis; if deeper the electrode will pierce the aponeurosis, and it will be in the second dorsal interosseus.

Third Lumbrical: If the electrode is inserted too deeply it will pierce the aponeurosis, and it will be in the second volar interosseus.

Fourth Lumbrical: If the electrode is inserted too deeply it will be in the opponens digiti minimi; if deeper the electrode will pierce the aponeurosis, and it will be in the third volar interosseus.

Comment

(a) Only 30 to 50 percent of hands have classically described innervation of first and second lumbricals being median innervated and the third and fourth being ulnar innervated.

(b) When classical innervation is present, median nerve entrapment or injury may result in involvement of the first and second lumbrical while ulnar nerve injury or entrapment may result in third and fourth lumbrical involvement.

(c) Involved in Klumpke's palsy.

FLEXOR POLLICIS BREVIS

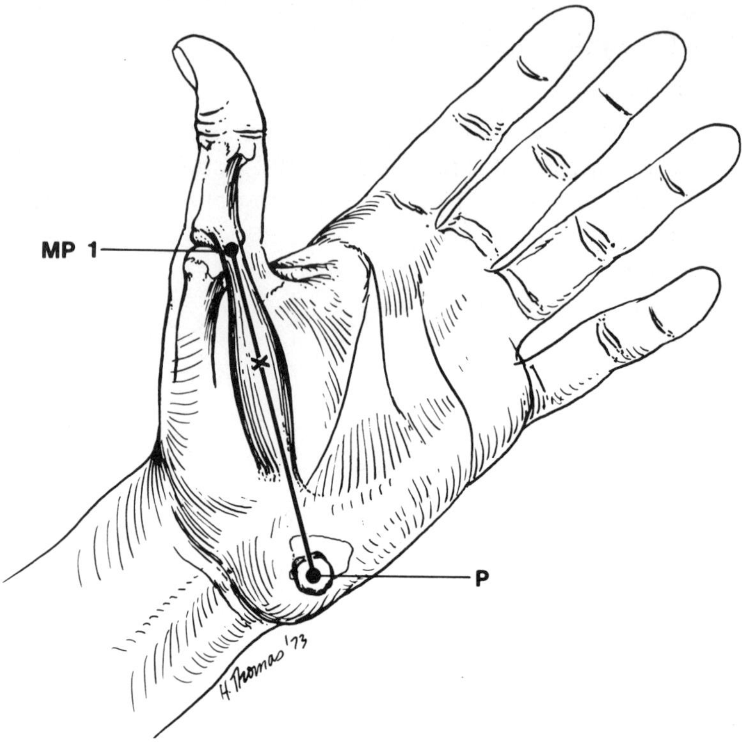

Innervation

Superficial Head: Median Nerve, Medial Cord, Anterior Division, Lower Trunk, *C8,* TI.

Deep Head: Ulnar Nerve, Medial Cord, Anterior Division, Lower Trunk, C8, *TI.*

Origin

Superficial Head: Ridge of the trapezium and the flexor retinaculum.

Deep Head: Ulnar side of first metacarpal.

Insertion

Superficial Head: Radial side of the base of the proximal phalanx of thumb.

Deep Head: Ulnar side of base of proximal phalanx of the thumb.

Position

Hand in full supination.

Electrode Insertion (X)

Superficial Head: A line is drawn between the ulnar aspect of the metacarpophalangeal joint (MP-I) and the pisiform (P). The needle is inserted at the junction between the middle and the radial third of this line to a depth of one-fourth to one-half inch.

Deep Head: As above but insert to depth of one-half to three-fourths inch.

Maneuver

Flex the metacarpophalangeal joint of the thumb.

Pitfalls

If the electrode is inserted too deeply it will be in the opponens pollicis, and if still deeper it will be in the adductor pollicis brevis.

Comment

(a) Two sesamoid bones are easily palpable in tendon at metacarpophalangeal joint.
(b) Due to insertion into extensor mechanism of thumb, it can extend IP joint of thumb when extensor pollicis longus is paralysed.
(c) Deep head involved in ulnar nerve injuries. Superficial head involved in ulnar nerve injuries.
(d) Involved in Klumpke's palsy (avulsion of C8, TI roots).

OPPONENS DIGITI MINIMI

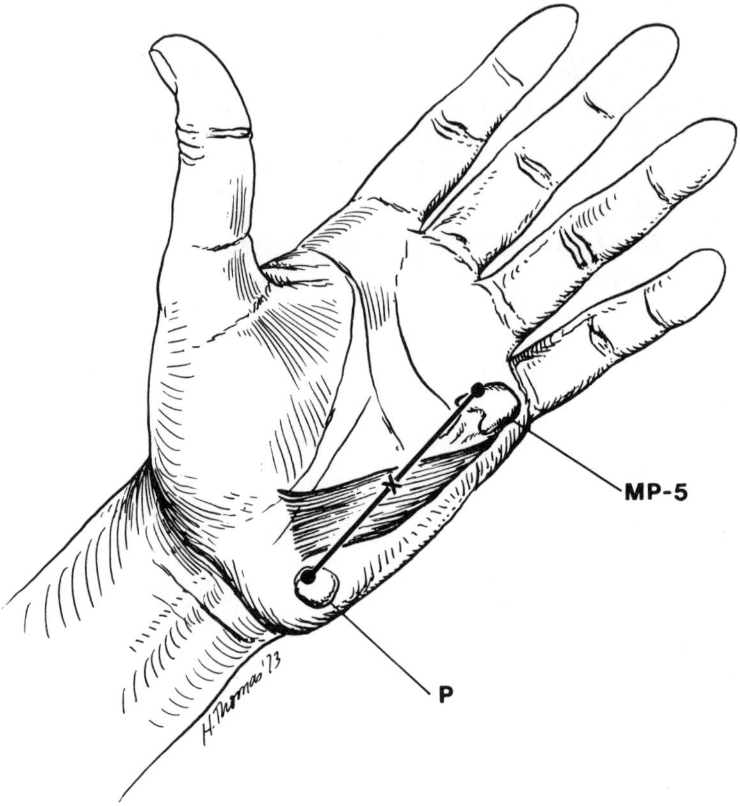

Innervation

Ulnar Nerve, Medial Cord, Anterior Division, Lower Trunk, C8, *TI.*

Origin

From the flexor retinaculum and the hook of the hamate.

Insertion

Into the medial surface of the fifth metacarpal.

Position

Hand in full supination.

Electrode Insertion (X)

Midpoint of a line drawn between the radial aspect of the fifth metacarpophalangeal joint (MP-5) and the radial aspect of the pisiform (P).

Test Maneuver

Oppose the little finger to the thumb.

Pitfalls

If the electrode is inserted too deeply it will be in fourth lumbrical; if deeper the palmar aponeurosis will be pierced, and the electrode will be in the third volar interosseus.

Comment

(a) Paralysis prevents full "cupping of hand."

(b) Involved in ulnar nerve injuries proximal to Guyon's Tunnel (cubital tunnel, Tardy palsy, Klumpke's palsy).

OPPONENS POLLICIS

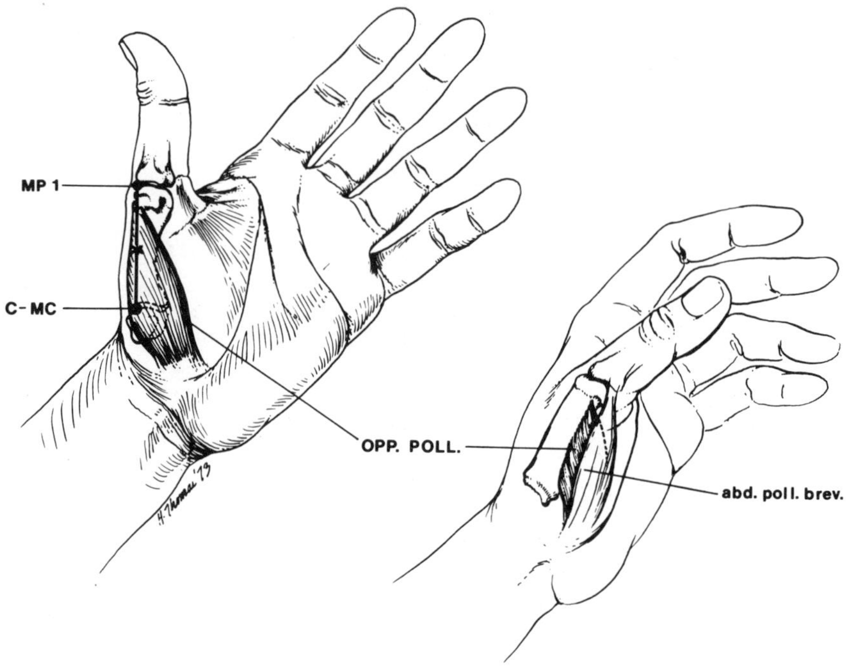

Innervation

Median Nerve, Medial Cord, Anterior Division, Lower Trunk, *C8,* TI.

Origin

From the tubercle of the trapezium and the flexor retinaculum.

Insertion

Into the lateral half of the palmar surface of the first metacarpal.

Position

Hand in full supination.

Electrode Insertion (X)

Midpoint of a line drawn between the radial aspect of the carpometacarpal (C-MC) and the metacarpophalangeal joints (MP-1). The electrode is placed between the abductor pollicis brevis and the first metacarpal to a depth of one-half to three-fourths inch.

Maneuver

Oppose thumb to little finger.

Pitfalls

If the electrode is inserted too deeply it will be in the adductor pollicis. If placed too medially it will be in the abductor pollicis brevis.

Comment

(1) The contribution to opposition made by this muscle is mainly rotation of the first metacarpal. The abductor pollicis brevis, and the adductor pollicis also contribute to the completion of this motion.
(2) May be involved in injuries or entrapment of median nerve in pronator muscle and carpal tunnel. Not involved in anterior interosseus entrapment or injury.
(3) Involved in Klumpke's palsy.

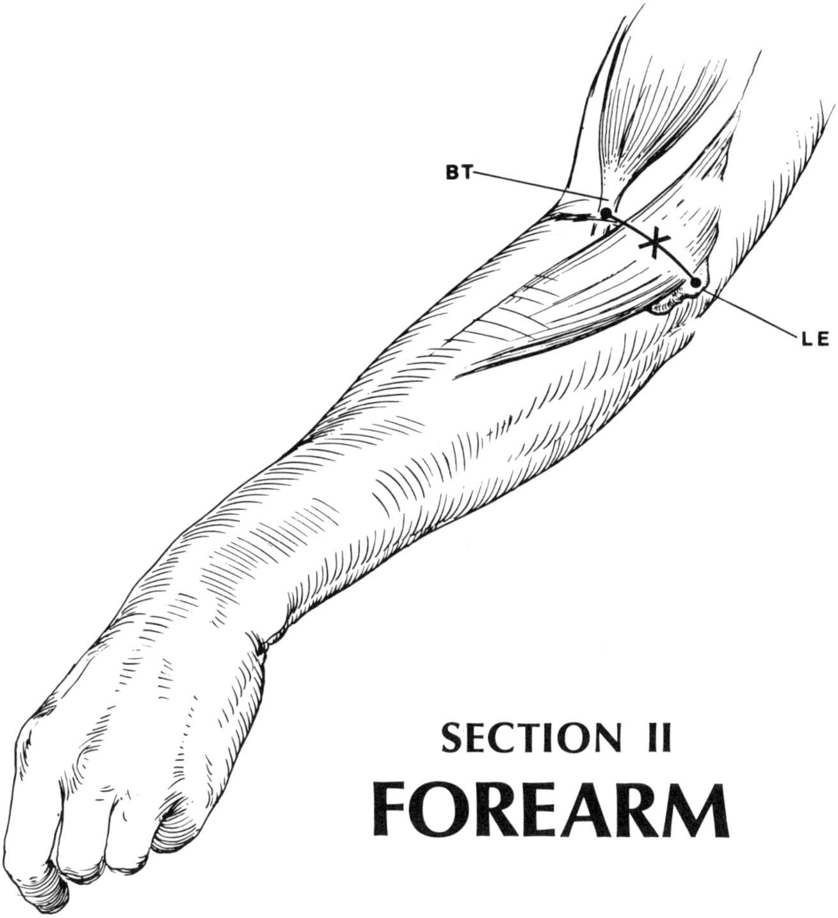

SECTION II
FOREARM

ABDUCTOR POLLICIS LONGUS

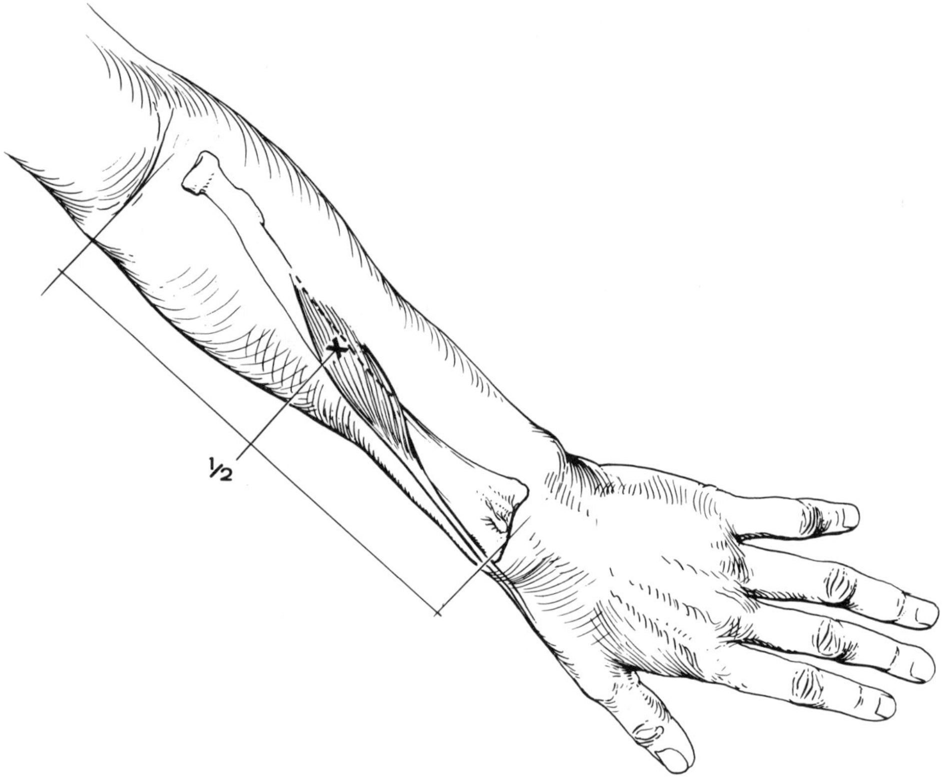

Innervation

Posterior Interosseus Nerve, Radial Nerve, Posterior Cord, Posterior Division, Middle and Lower Trunk, *C7,* C8.

Origin

From the dorsal surface of the body of the ulna, the interosseus membrane, and the middle one-third of the body of the radius.

Insertion

Lateral aspect of the base of the first metacarpal.

Position

Forearm fully pronated.

Electrode Insertion (X)

Over the shaft of the radius at mid-forearm.

Test Maneuver

Radial abduction of the thumb.

Pitfalls

If the electrode is inserted too proximally it will be in the extensor carpi radialis brevis; if inserted too distally it will be in the extensor pollicis brevis; if it is inserted too ulnarly it will be in the extensor digitorum communis.

Comment

(a) Tendon involved in DeQuervain's stenosing synovitis.
(b) Involved in posterior interosseus entrapment and more proximal injuries to the radial nerve.
(c) Tendon runs through the first compartment on dorsum of the wrist.

ANCONEUS

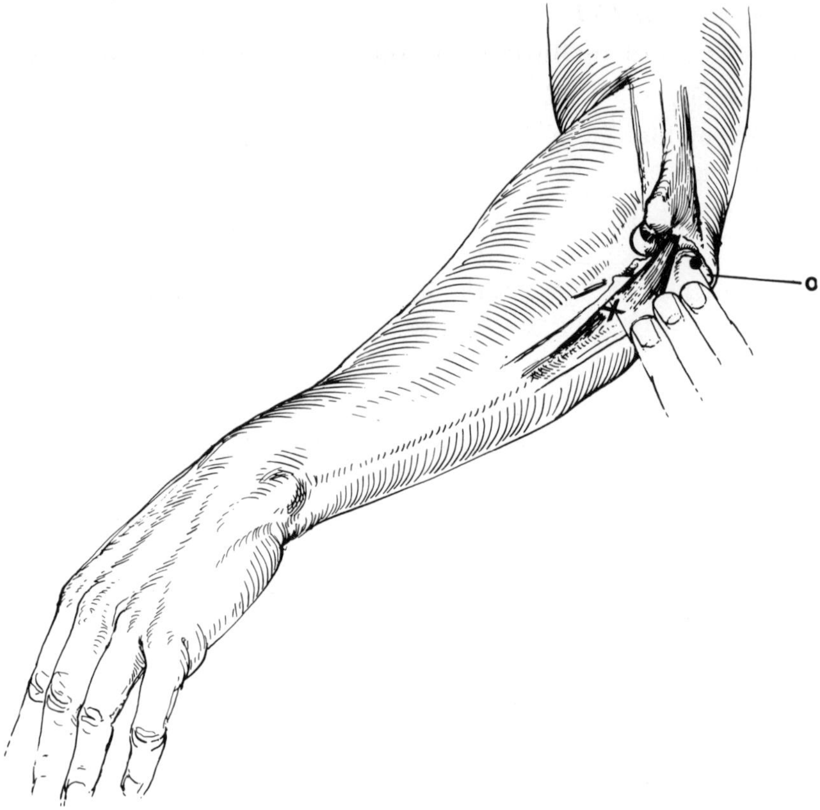

Innervation

Radial Nerve, Posterior Cord, Posterior Division, Middle and Lower Trunk, *C7,* C8.

Origin

From the posterior aspect of the lateral epicondyle of the humerus as a continuation of the medial head of the triceps.

Insertion

Lateral aspect of the olecranon process and the proximal portion of the posterior surface of the ulna.

Position

The forearm fully pronated and the elbow at ninety degrees of flexion.

Electrode Insertion (X)

Place the tip of the little finger on the olecranon (O) and ring and middle fingers along with ulna. Insert the electrode just beyond tip of middle finger, just radial to the ulna.

Test Maneuver

Extension of the elbow.

Pitfalls

If the needle is inserted too radially it will be in the extensor carpi ulnaris; if inserted too deeply it will be in the supinator.

Comment

This muscle may be considered a continuation of the medial head of the triceps to the lateral epicondyle. It is innervated by a long branch of the radial nerve, which results in its being spared except in very proximal injuries of the radial nerve.

BRACHIORADIALIS

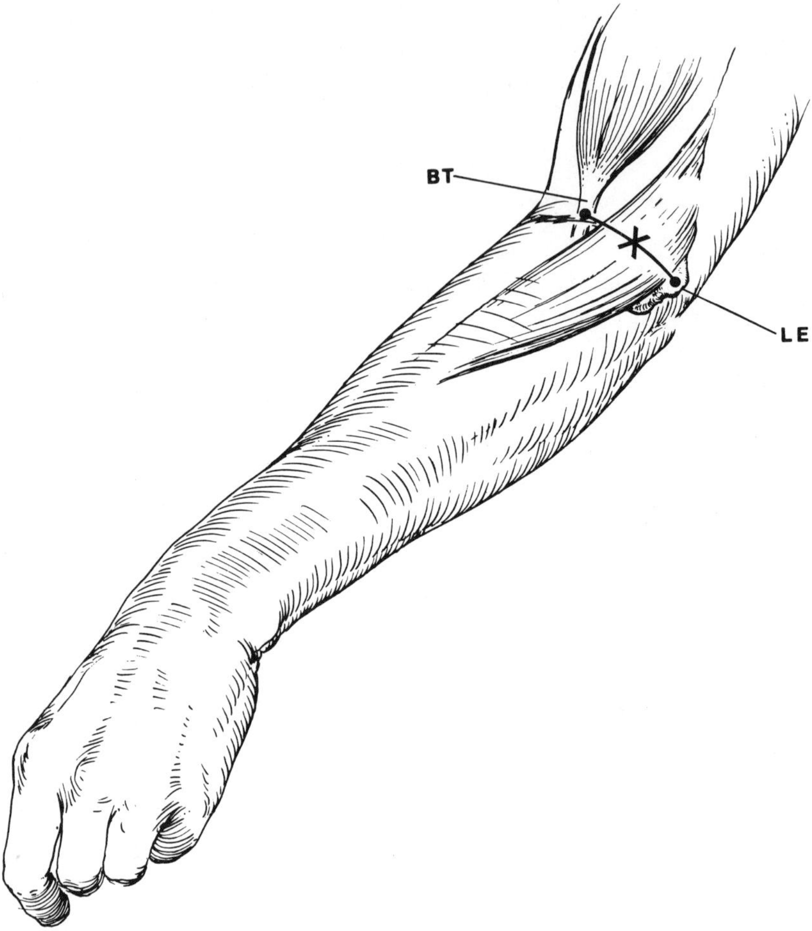

Innervation

Radial Nerve, Posterior Cord, Posterior Division, Upper Trunk C5, *C6.*

Origin

From the supracondylar area of the lateral aspect of the humerus.

Insertion

Lateral aspect of the radius, just above the styloid process.

Position

Forearm fully pronated.

Electrode Insertion (X)

Midway between biceps tendon (BT) and lateral epicondyle (LE) along flexor crease; insert electrode to a depth of one-half inch.

Test Maneuver

Flexion of the forearm in neutral position.

Pitfalls

If the needle is inserted too laterally it will be in the extensor carpi radialis longus.

Comment

(a) The only muscle producing flexion of the elbow supplied by radial nerve.
(b) Can act as supinator or pronator from the extremes of these positions, bringing the forearm into the neutral position.
(c) The only primary elbow flexor not supplied by the musculocutaneus nerve.
(d) Paralysed in radial nerve injuries above or at spiral groove of humerus.

EXTENSOR CARPI RADIALIS, LONGUS AND BREVIS[1]

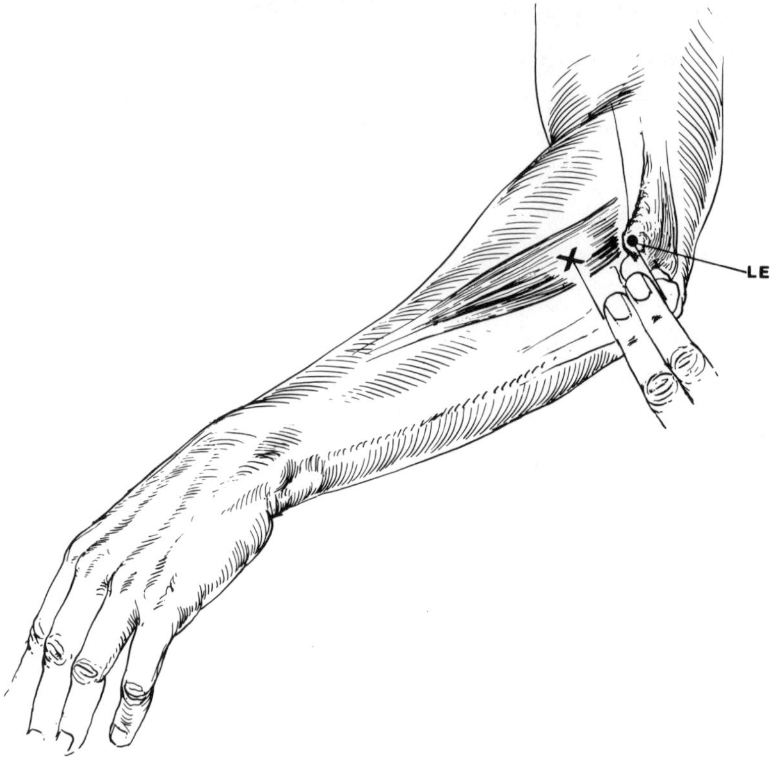

Innervation

Radial Nerve, Posterior Cord, Posterior Division, Upper and Middle Trunk, *C6,* C7.

Origin

Longus: The lower third of the supracondylar ridge of humerus.

Brevis: The lateral epicondyle of humerus.

Insertion

Longus: Dorsal surface of base of second metacarpal.

Brevis: Dorsal surface of third metacarpal.

---

[1] Because of the close anatomical and functional relationship of these muscles, the authors have found it impossible to develop a technique by which they could place the electrode in one or the other of these muscles with confidence.

Position

Forearm fully pronated.

Electrode Insertion (X)

Insert two fingerbreadths distal to lateral epicondyle (LE).

Test Maneuver

Dorsiflexion of wrist in radial deviation.

Comment

(a) The tendons of these muscles occupy the second extensor compartment on dorsum of the wrist.
(b) Usually spared in posterior interosseus syndrome but usually involved in lesions at or above the spiral groove (of humerus).
(c) Frequently involved in "Saturday night palsy."

EXTENSOR CARPI ULNARIS

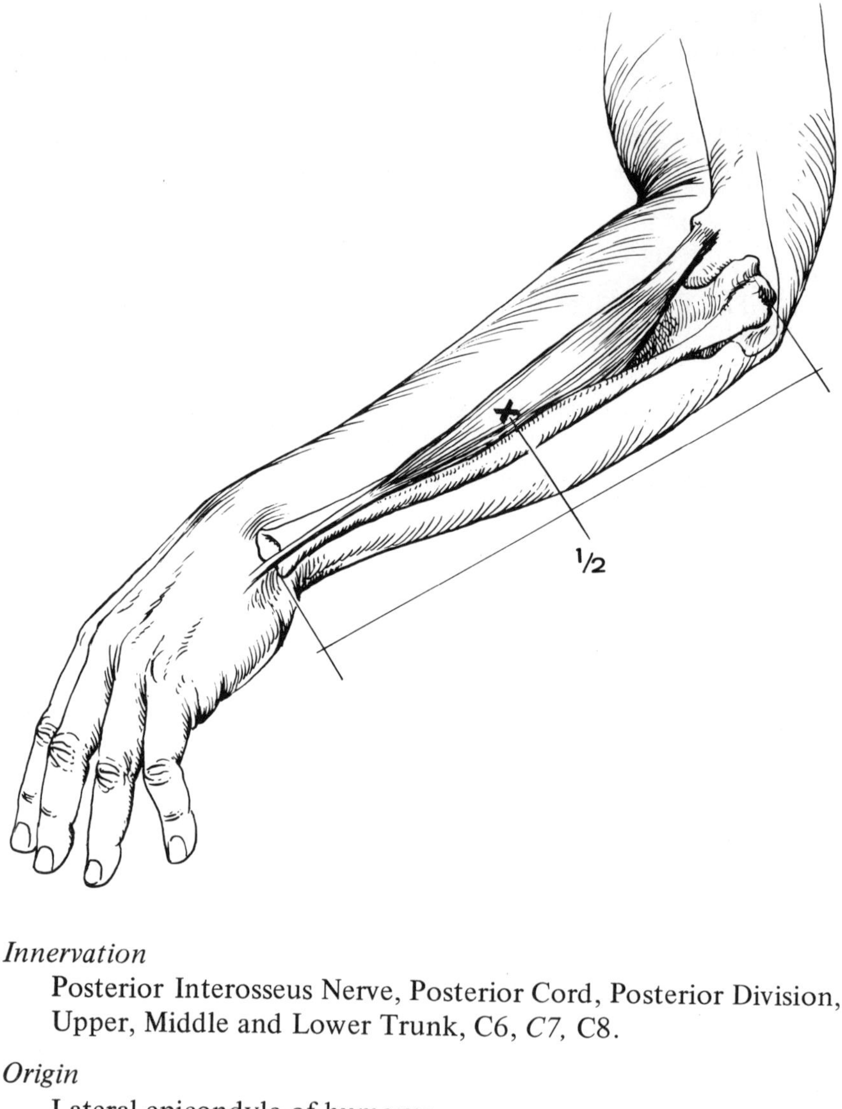

Innervation

Posterior Interosseus Nerve, Posterior Cord, Posterior Division, Upper, Middle and Lower Trunk, C6, *C7,* C8.

Origin

Lateral epicondyle of humerus.

Insertion

Dorsal surface of base of fifth metacarpal.

Position

The forearm fully pronated.

Electrode Insertion (X)

Palpate the ulna in middle of forearm and insert needle electrode just above the shaft of ulna.

Test Maneuver

Extend the wrist with ulnar deviation.

Pitfalls

If the needle electrode is inserted too radially it will be in the extensor pollicis longus, and if inserted too proximally it will be in the anconeus.

Comment

(a) The tendon of this muscle occupies the sixth extensor compartment on dorsum of wrist.

(b) Involved in posterior interosseus nerve lesions, Saturday night palsy and crutch palsy.

EXTENSOR DIGITORUM COMMUNIS AND EXTENSOR DIGITI QUINTI PROPRIUS[2]

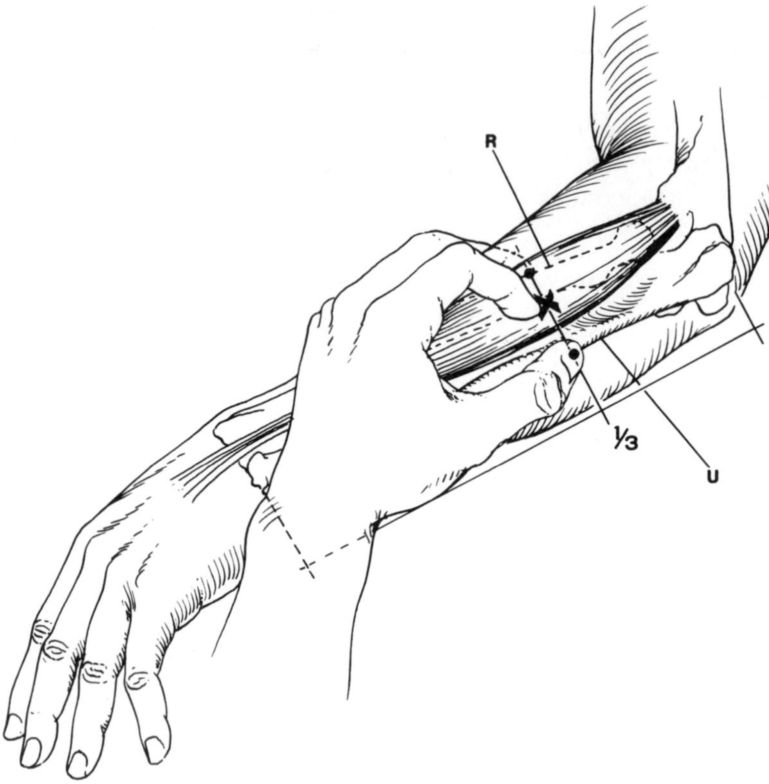

Innervation

Posterior Interosseus Nerve, Radial Nerve, Posterior Cord, Posterior Division, Middle and Lower Trunk, *C7,* C8.

Origin

Common extensor tendon from the lateral epicondyle of humerus.

[2] Extensor Digiti Quinti Porprius: Is most frequently a part of the extensor digitorum communis. This muscle lies very close to the extensor digitorum communis but arises from a separate slip and can be found in the mid-forearm on the ulnar border of the extensor digitorum communis.

Insertion

On dorsal surface of base of second to fifth phalanges of fingers.

Position

The forearm fully pronated.

Electrode Insertion (X)

Grasp the forearm at function of upper and middle third with thumb and middle finger on radius (R) and ulna (U). Then with index finger bissect these two points and insert needle electrode at tip of index finger to a depth of one-half inch.

Test Maneuver

Extend metacarpophalangeal joints.

Pitfalls

If the needle electrode is inserted too deeply it will be in the extensor pollicis longus; if inserted too medially it will be in the extensor carpi radialis brevis; if inserted too laterally it will be in the extensor carpi ulnaris.

Comment

Involved in posterior interosseus and more proximal radial nerve lesion.

(a) The tendon of the extensor digitorum communis occupies the fourth extensor compartment on the dorsum of the wrist while the tendon of the extensor digit quinti proprius goes throughout the fifth compartment.
(b) Involved in lesions of posterior interosseus nerve lesions, "Saturday night" palsy and crutch paralysis.

EXTENSOR INDICIS PROPRIUS

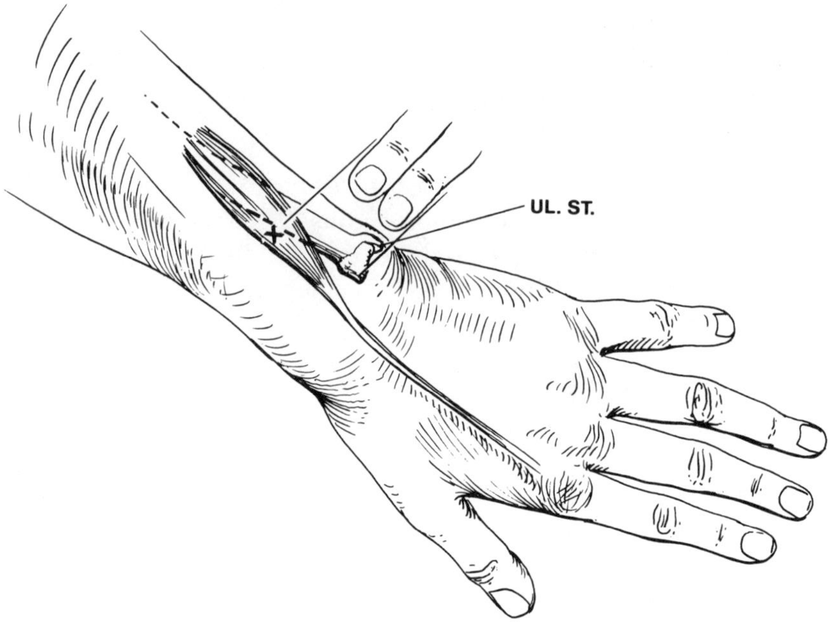

Innervation

Posterior Interosseus Nerve, Radial Nerve, Posterior Cord, Posterior Division, Middle and Lower Trunk, *C7,* C8.

Origin

Dorsal surface of lower half of ulnar shaft below the origin of the extensor pollicis longus.

Insertion

Joins ulnar side of tendon of extensor digitorum communis, which goes to index finger; terminates in extensor expansion.

Position

The forearm fully pronated.

Electrode Insertion (X)

Two fingerbreadths proximal to ulnar styloid (UL.ST.) just radial to ulnar at a depth of one-half inch.

Test Maneuver

Extend finger with flexion of other fingers.

Pitfalls

If needle electrode is inserted too radially it will be in the abductor pollicis longus; if inserted too proximally it will be in the extensor digitorum communis.

Comment

(a) Usually the most distal radial nerve innervated muscle (at times the extensor pollicis longus occupies this position).
(b) Tendon occupies the fourth compartment on dorsum of wrist with extensor digitorum communis.
(c) Used as recording muscle in radial nerve motor conduction studies.
(d) Involved in posterior interosseus and more proximal radial nerve injuries ("Saturday night" palsy and crutch palsy).

EXTENSOR POLLICIS BREVIS

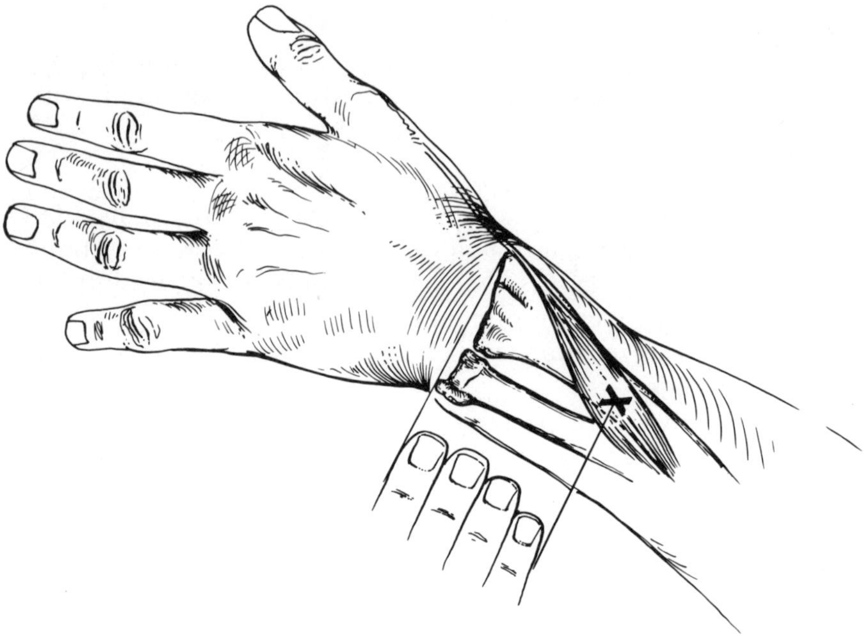

Innervation

Posterior Interosseus, Radial Nerve, Posterior Cord, Posterior Division, Middle and Lower Trunk, *C7,* C8.

Origin

Dorsal surface of radial shaft below abductor pollicis longus.

Insertion

The dorsal aspect of the first phalanx of thumb.

Position

The forearm fully pronated.

Electrode Insertion (X)

Insert needle electrode directly over the ulnar side of radius, four fingerbreadths proximal to wrist.

Test Maneuver

Extend proximal phalanx of thumb.

Pitfalls

If the needle electrode is inserted too proximally it will be in the abductor pollicis longus.

Comment

(a) Tendon runs through the first compartment on dorsum of wrist. Distal to this compartment, the tendon forms the radial border of the anatomical "Snuffbox."

(b) Most distal muscle innervated by radial nerve through posterior interosseus branch.

(c) Involved in lesions of posterior interosseus and more proximal radial nerve lesions.

EXTENSOR POLLICIS LONGUS

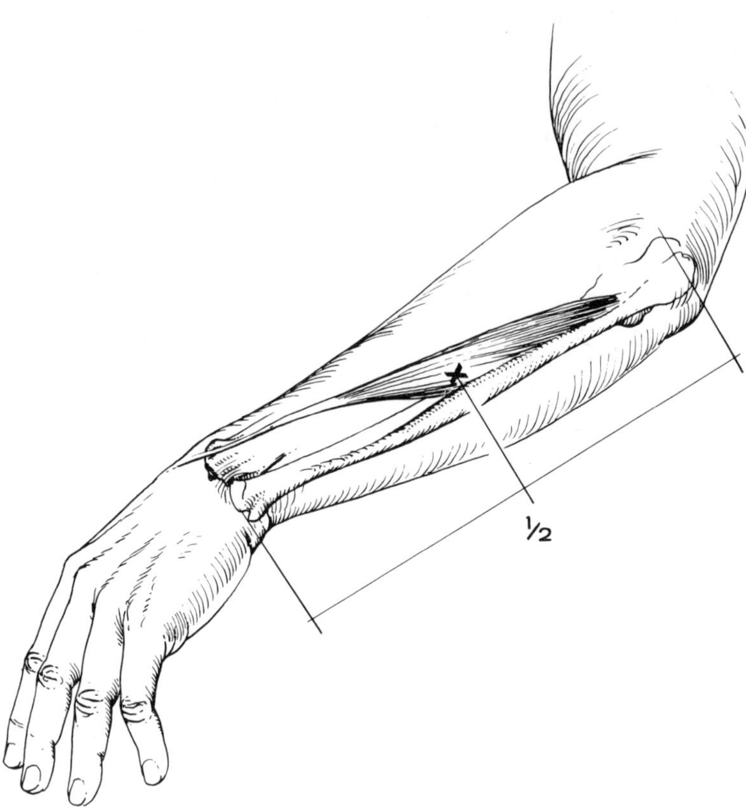

Innervation

Posterior Interosseus Nerve, Radial Nerve, Posterior Cord, Posterior Division, Middle and Lower Trunk, *C7*, C8.

Origin

The dorsal surface of the middle third of ulnar shaft below the abductor pollicis longus.

Insertion

The dorsal aspect of base of terminal phalanx of thumb.

Position

The forearm fully pronated.

Electrode Insertion (X)

At mid-forearm insert needle electrode along radial border of ulnar.

Test Maneuver

Extend distal phalanx of thumb.

Pitfalls

If the needle electrode is inserted too ulnarly it will be in the extensor digitorum communis; if inserted too proximally it will be in the abductor pollicis longus.

Comment

(1) Involved in posterior interosseus and more proximal radial nerve lesions.

(2) When paralysed, the flexor pollicis brevis may extend terminal phalanx weakly and give false impression of intact extensor pollicis longus.

(3) The tendon of this muscle occupies the third extensor compartment on the dorsum of the wrist. Distal to this compartment the tendon forms the ulnar border of the "Snuffbox."

FLEXOR CARPI RADIALIS

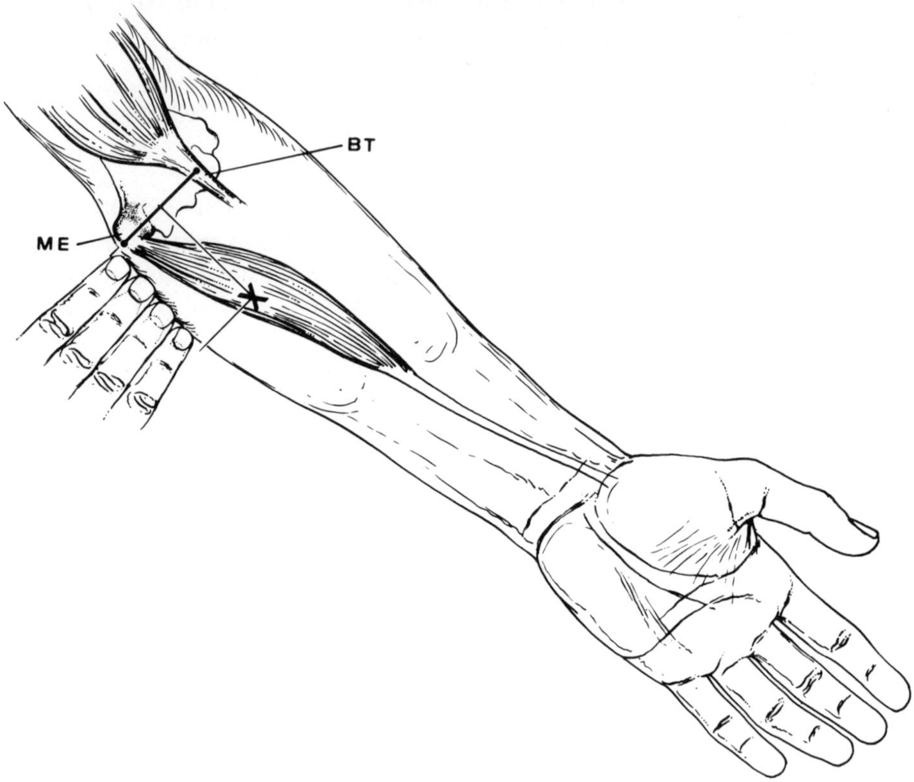

Innervation

Median Nerve, Lateral and Medial Cord, Anterior, Divisions, Upper, Middle and Lower Trunk, C6, *C7,* C8.

Origin

Common tendon from medial epicondyle of humerus.

Insertion

Volar surface of base of second metacarpal.

Position

The forearm fully supinated.

Electrode Insertion (X)

Three to four fingerbreadths distal to the midpoint of a line connecting the medial epicondyle (ME) and biceps tendon (BT).

Test Maneuver

Flexion of wrist with radial deviation.

Pitfalls

If the needle electrode is inserted too deeply it will be in the flexor digitorum sublimus; if deeper it will be in the flexor pollicis longus. If inserted too laterally it will be in the pronator teres, and if inserted too medially it will be in the palmaris longus.

Comment

(a) The median nerve runs just ulnar to its tendon as it crosses the volar aspect of the wrist.

(b) Together with the flexor pollicis longus and the radial artery they form the "Radial Trio."

FLEXOR CARPI ULNARIS

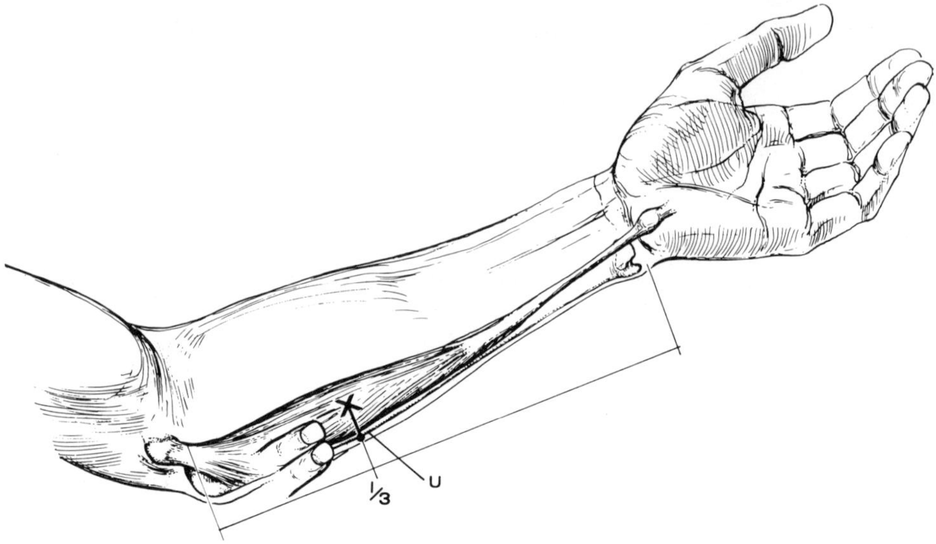

Innervation

Ulnar Nerve, Medial Cord, Anterior Division, Lower Trunk, *C8,* TI.

Origin

Common tendon from medial epicondyle of humerus, medial margin of olecranon, coronoid process, and upper two-thirds of the dorsal border of ulna.

Insertion

Volar surface of pisiform, hamate and fifth metacarpal.

Position

The forearm fully supinated.

Electrode Insertion (X)

Two fingerbreadths volar to ulna (U) at the junction of the upper and middle thirds of the forearm.

Test Maneuver

Flexion of the wrist with ulnar deviation.

Pitfalls

If the needle electrode is inserted too deeply it will be in the flexor digitorum profundus.

Comment

(a) The ulnar nerve is located just radial to its tendon at the wrist.

(b) Involved in lesions of ulnar nerve at or above ulnar groove.

(c) Together with the ulnar nerve and the ulnar artery they form the "ulnar trio."

FLEXOR DIGITORUM PROFUNDUS

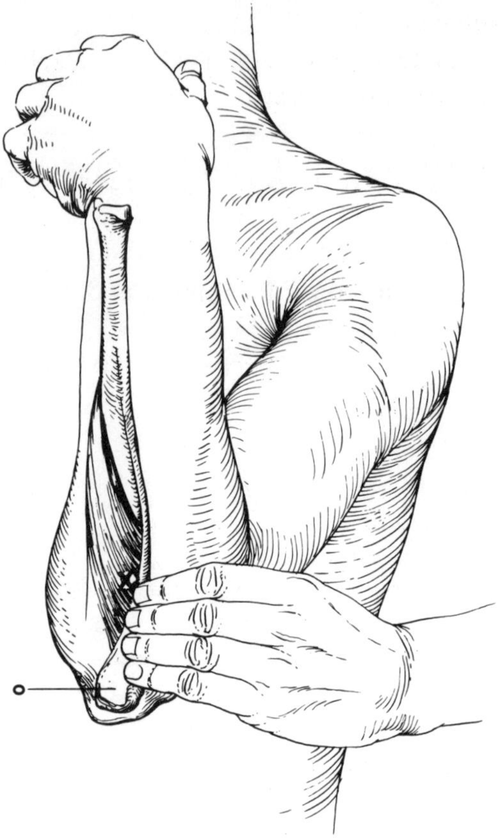

Innervation

Digits Two and Three: Anterior Interosseus Nerve, Median Nerve, Medial Cord, Anterior Division, Middle and Lower Trunk, C7, *C8*.

Digits Four and Five: Ulnar Nerve, Medial Cord, Anterior Division, Lower Trunk, *C8,* TI.

Origin

Upper three-fourths of volar and medial surfaces of ulna.

Insertion

Volar surfaces of bases of distal phalanges of four digits.

Position

The forearm fully supinated.

Electrode Insertion (X)

Place tip of little finger on olecranon (O) and ring, middle and index fingers along shaft of ulnar. Insert needle electrode just beyond tip of index finger just ulnarly to shaft. The ulnar innervated portion is the more superficial (1-2 cm), while the median innervated portion is deeper (3-5 cm).

Test Maneuver

Flexion of distal phalanges of digits.

Pitfalls

If the needle electrode is inserted too volarly it will be in the flexor carpi ulnaris.

Comment

(a) The four tendons run across the wrist resting on the pronator quadratus. They are deep to the flexor sublimis tendon and the median nerve.
(b) The two median innervated heads are involved in anterior interosseus nerve, pronator and ligament of Struther entrapments manifested by weakness or inability to flex the terminal phalanges of second and third digits.
(c) The two ulnar innervated heads are involved in cubital tunnel entrapment and higher ulnar, medial cord and C8, TI lesions. This is manifested by weakness or mobility to flex the fourth and fifth terminal phalanges.

FLEXOR DIGITORUM SUBLIMIS

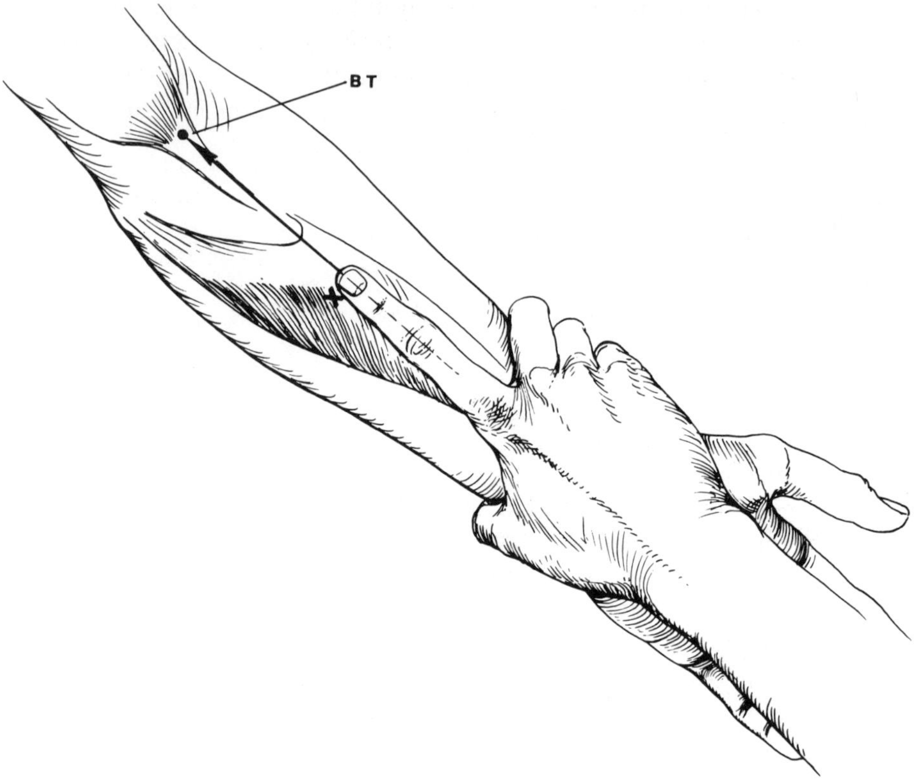

Innervation

Median Nerve, Lateral and Medial Cord, Anterior Division, Middle and *Lower Trunk,* C7, *C8,* TI.

Origin

Common tendon from medial epicondyle of humerus; coronoid process of ulnar and oblique line of radius.

Insertion

Volar surfaces of bases of second phalanges of digits.

Position

The forearm fully supinated.

Electrode Insertion (X)

Grasp with operator's palm to volar surface of subject's wrist. Point index finger to biceps tendon (BT) and insert needle electrode just ulnarly to tip of index finger.

Test Maneuver

With the distal interphalangeal joints of three digits maintained in hyper extension, the patient is asked to flex the proximal interphalangeal joint of the free finger.

Pitfalls

If the needle electrode is inserted too radially it will be in the flexor carpi radialis; if inserted too ulnarly it will be in the flexor digitorum profundus; if too distally it will be in the tendon of the flexor carpi radialis longus.

Comment

(a) The four tendons run across the wrist in pairs: the tendons for the middle and ring fingers are the most superficial; those to the ring and little finger are deeper resting on the flexor digitorum profundus tendons. The medial nerve lies radial to these tendons as it enters the carpal tunnel.

(b) Involved in median nerve entrapments (pronator teres; ligament of Struther) and higher median nerve lesions.

FLEXOR POLLICIS LONGUS

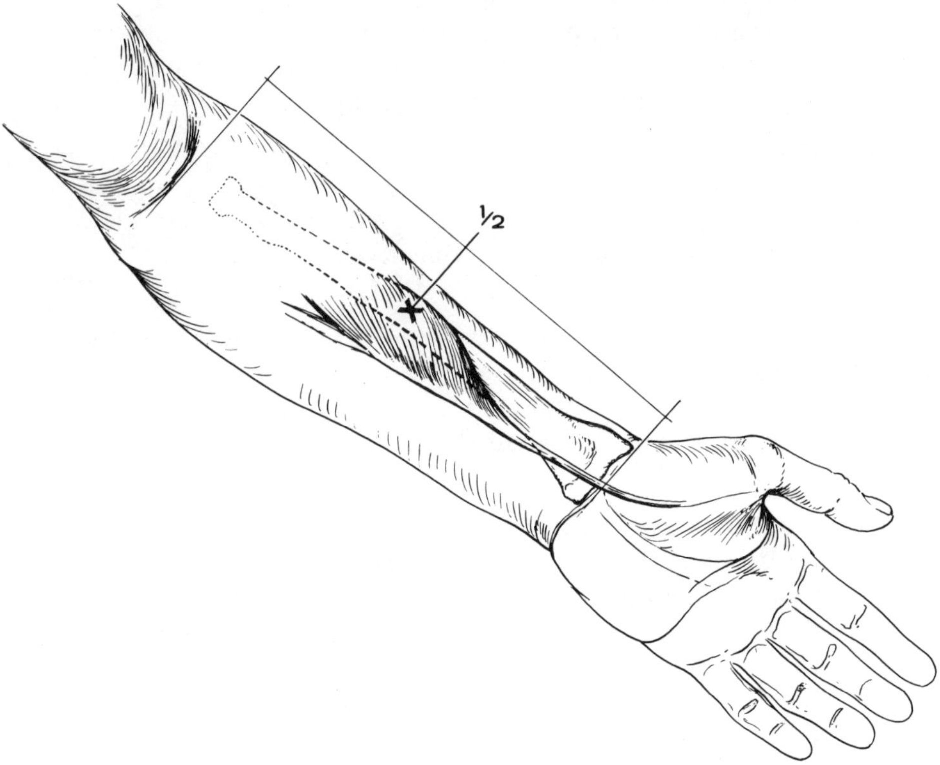

Innervation

Anterior Interosseus Nerve, Median Nerve, Lateral and Medial Cords, Anterior Divisions, Middle and Lower Trunks, C7, *C8,* TI.

Origin

The volar surface of the body of the radius from the bicipital tuberosity to the attachment of the pronator quadratus and interosseus membrane.

Insertion

Volar surface of base of distal phalanx of thumb.

Position

The forearm fully supinated.

Electrode Insertion (X)

In the middle of the forearm the needle electrode is inserted from the radial aspect just volar to the radius.

Test Maneuver

Flexion of the distal phalanx of thumb.

Pitfalls

If the needle electrode is inserted too superficially it will be in the flexor digitorum sublimis.

Comment

(a) The tendon of this muscle runs deep across the wrist, radial to the flexor digitorum profundus. With the flexor carpi radialis tendon and the radial artery, it forms the "Radial Trio."

(b) In patients with ulnar nerve lesions with paralysis of adductor pollicis, the flexor pollicis longus is substituted, producing flexion of the terminal phalanx of the thumb (Froment's "signe de journal").

(c) Involved in anterior interosseus, pronator teres and ligament of Struther entrapments.

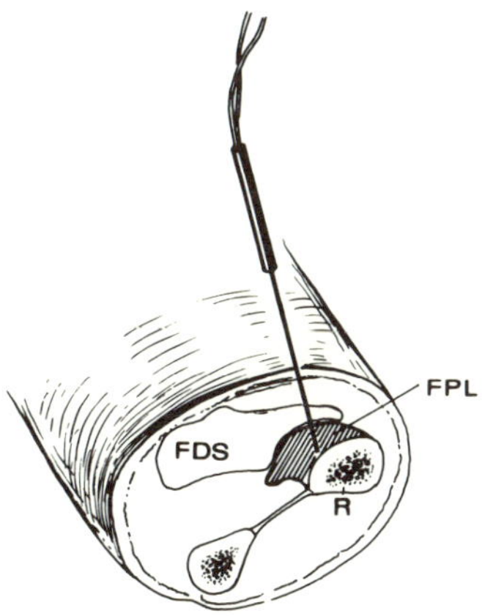

FPL–Flexor Pollicis Longus
FDS-Flexor Digitorum Sublimis
R–Radius

PALMARIS LONGUS

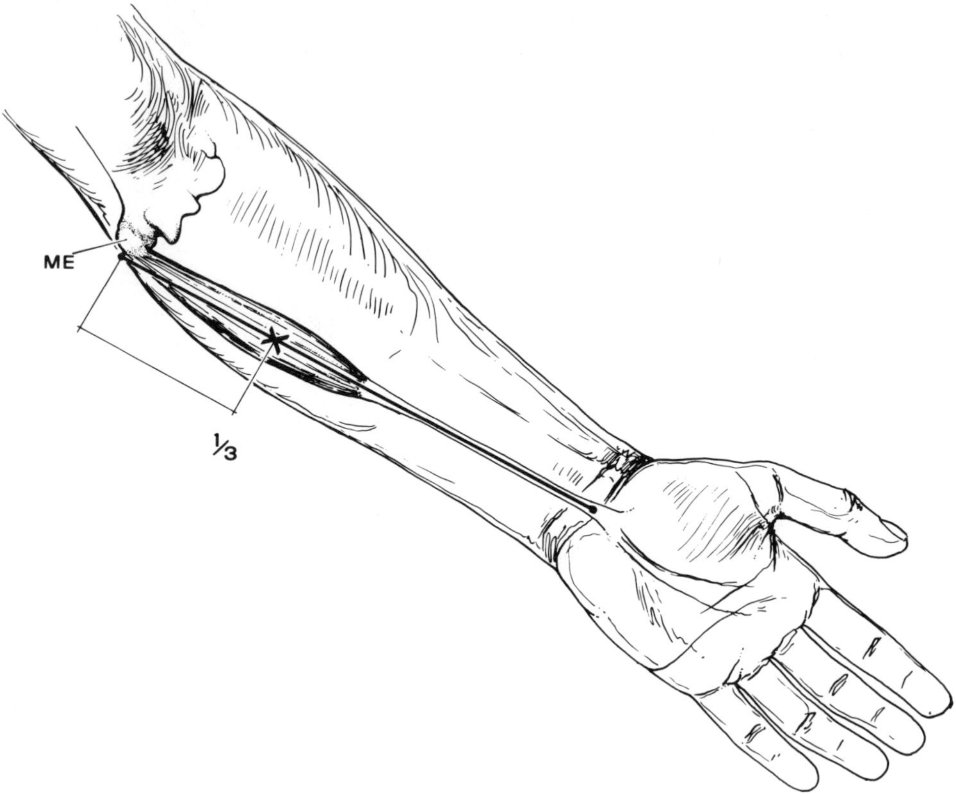

Innervation

Median Nerve, Lateral and Medial Cord, Anterior Division, Middle and *Lower* Cord, C7, *C8,* TI.

Origin

The medial epicondyle of humerus.

Insertion

The palmar aponeurosis.

Position

The forearm fully supinated.

Electrode Insertion (X)

At the junction of the upper and middle thirds of a line joining the medial epicondyle (ME) and middle of volar surface of wrist.

Test Maneuver

Cup palm of hand.

Pitfalls

If the needle electrode is inserted too medially it will be in the flexor carpi ulnaris; if inserted too radially it will be in the flexor carpi radialis, and if inserted too deeply it will be in the flexor digitorum sublimus.

Comment

(a) Absent in 10 to 15 percent of individuals.
(b) Only flexor tendon superficial to the volar carpal ligament.
(c) Involved in median nerve entrapment syndromes (pronator teres, ligament of Struther).

PRONATOR QUADRATUS

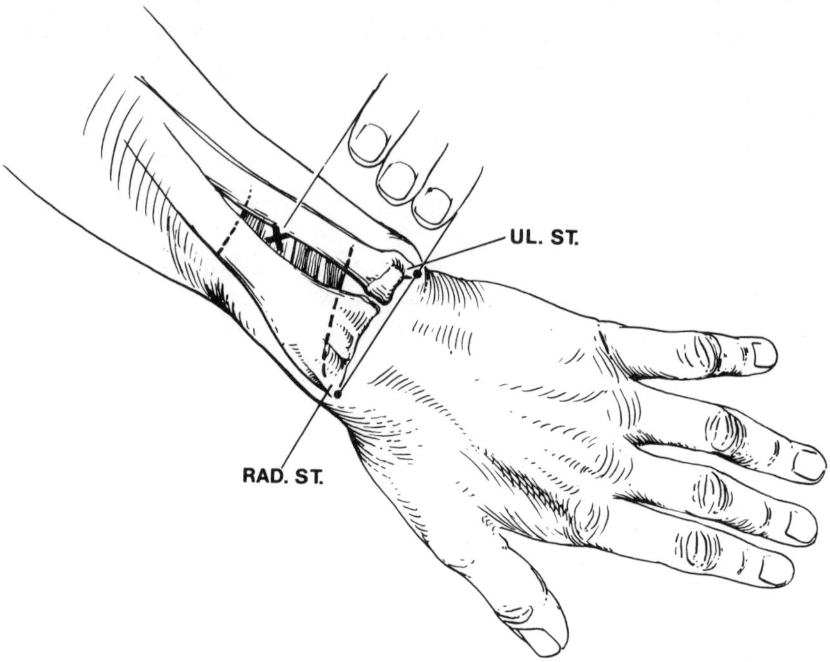

Innervations

Anterior Interosseus Nerve, Median Nerve, Lateral and Medial cords, Anterior Divisions, Middle and Lower Trunks, C7, *C8*, TI.

Origin

Lower fourth of volar surface of ulnar.

Insertion

Lower fourth of lateral border and volar surface of shaft of radius.

Position

The forearm fully pronated.

Electrode Insertion (X)

Three fingerbreadths proximal to midpoint of a line connecting the radial (RAD.ST.) and ulnar styloids (UL.ST.); insert needle electrode through the interosseus membrane to a depth of approximately three-quarter inches.

Test Maneuver

Pronation of forearm.

Pitfalls

If the needle electrode is inserted too deeply, it will be in the flexor digitorum sublimis.

Comment

(a) The most distal muscle innervated by the anterior interosseus nerve.

(b) Involved in median nerve entrapment syndromes (anterior interosseus, pronator teres; ligament of Struther).

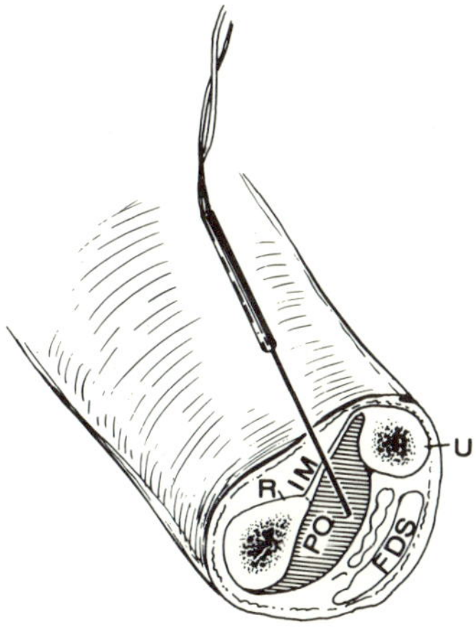

PQ–Pronator Quadratus
FDS–Flexor Digitorum Sublimis
R–Radius
IM–Interosseus Membrane
U–Ulnar

PRONATOR TERES

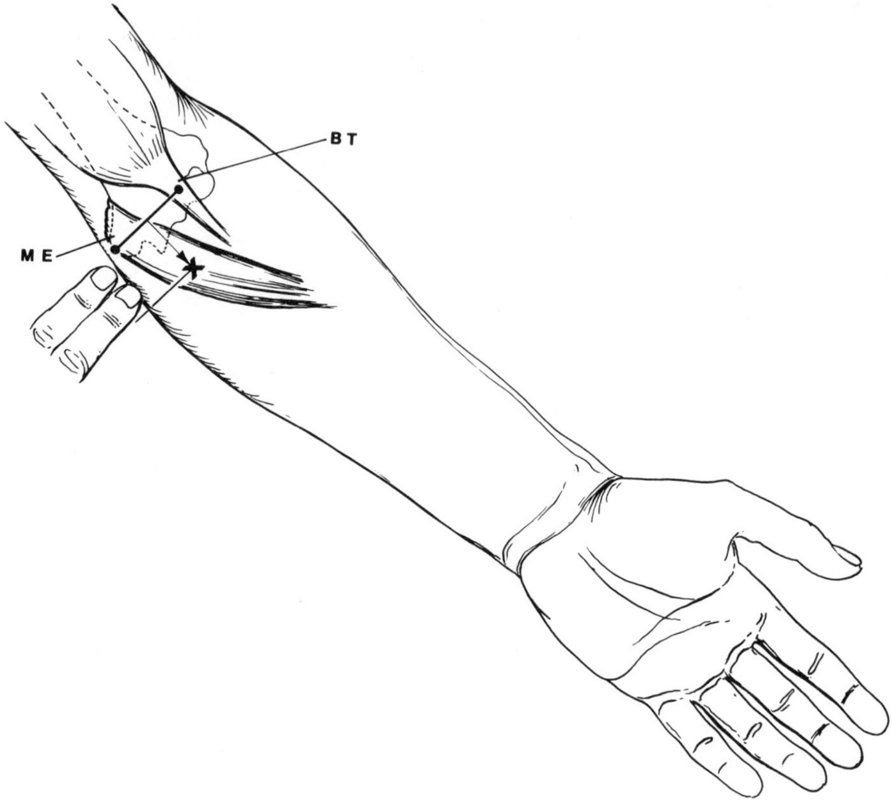

Innervation

Median Nerve, Lateral Cord, Anterior Division, Upper and *Middle* Trunk, C6, *C7.*

Origin

Common tendon from medial epicondyle of humerus and coronoid process of ulnar.

Insertion

Lateral surface of radius at mid-shaft.

Position

The forearm fully supinated.

Electrode Insertion (X)

Two fingerbreadths distal to the midpoint of a line connecting the medial epicondyle (ME) and biceps tendon. (BT)

Test Maneuver

Pronation of forearm.

Pitfalls

If the needle electrode is inserted too deeply it will be in the flexor digitorum sublimis; if inserted too ulnarly it will be in the flexor carpi radialis.

Comment

(a) The most proximal muscle innervated by the median nerve.
(b) Common site of entrapment as it is pierced by the median nerve.
(c) May or may not be involved in pronator teres syndrome depending on whether the nerve to the pronator muscle branches proximal to or within the muscle itself.
(d) Also involved in entrapment at the ligament of Struther.

SUPINATOR

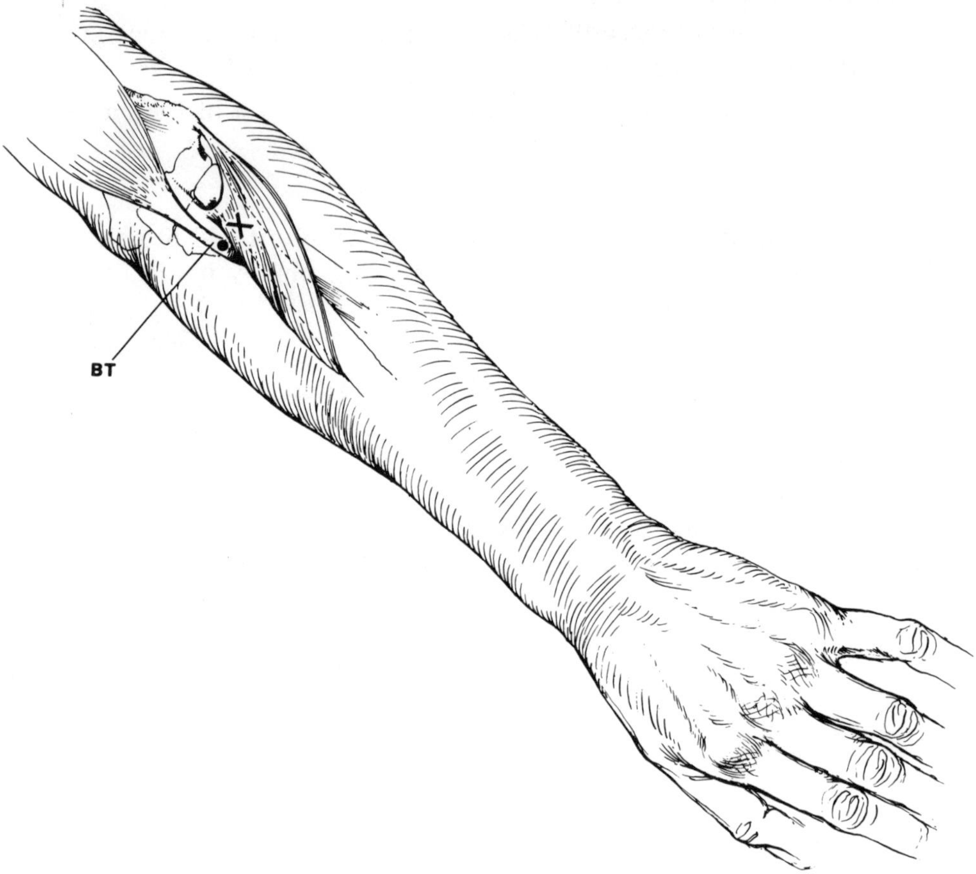

Innervation

Posterior Interosseus Nerve, Radial Nerve, Posterior Cord, Posterior Division, Upper Trunk, C5, *C6.*

Origin

Lateral epicondyle of humerus, radial collateral ligament of elbow, supinator crest of ulna.

Insertion

Dorsal and lateral surfaces of upper third of radial shaft.

Position

The forearm fully pronated.

Electrode Insertion (X)

Just radial to the most distal part of insertion of the biceps tendon (BT).

Test Maneuver

Supination of forearm.

Pitfalls

If needle electrode is inserted too laterally it will be in the brachioradialis. There is danger of puncturing the radial artery.

Comment

The posterior interosseus nerve passes through an aponeurotic arch (the Arcade of Fröhse) between the two heads of this muscle. When this becomes thickened and tight, it might entrap the nerve.

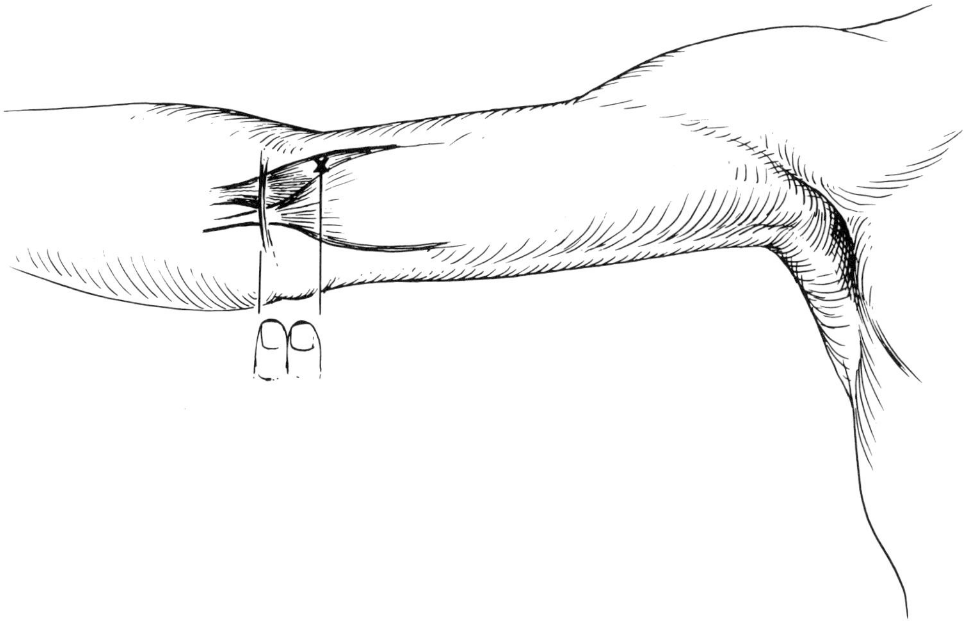

SECTION III
ARM

BICEPS BRACHII

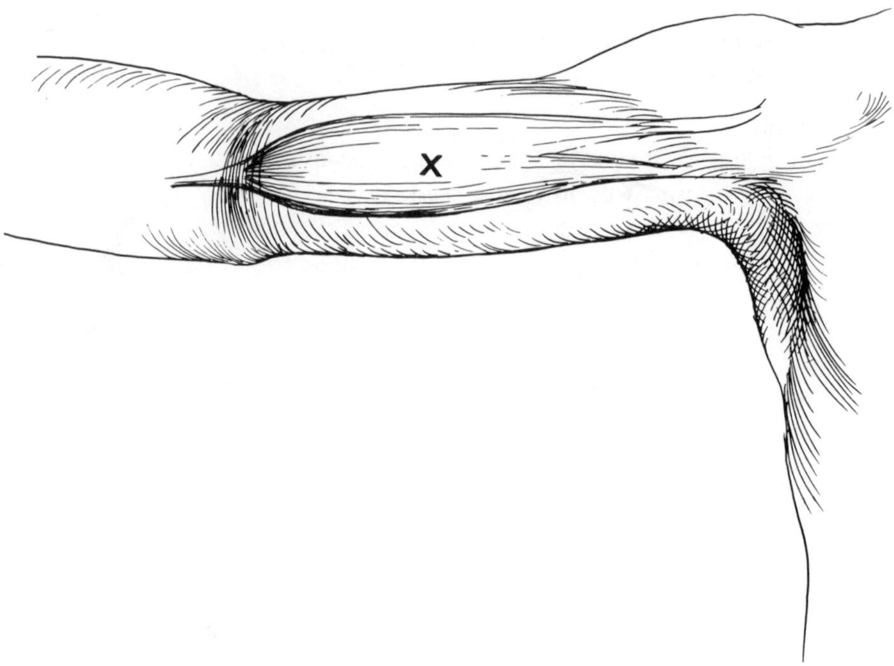

Innervation

Musculocutaneous Nerve, Lateral Cord, Anterior Division, Upper Trunk, *C5, C6.*

Origin

Long Head: From the supraglenoid tuberosity of scapula.

Short Head: From the apex of the coracoid process of the scapula.

Insertion

On the bicipital tuberosity of the radius.

Position

The patient supine with the arm extended.

Electrode Insertion (X)

Into the bulk of the muscle in mid-arm.

Test Maneuver

Flex or supinate forearm.

Pitfalls

If the needle electrode is inserted too deeply it will be in the brachialis.

Comment

(a) Frequently used as recording muscle for musculocutaneus nerve motor conduction study.

(b) Involved in entrapment of musculocutaneus nerve as it courses through the corocobrachialis muscle.

BRACHIALIS

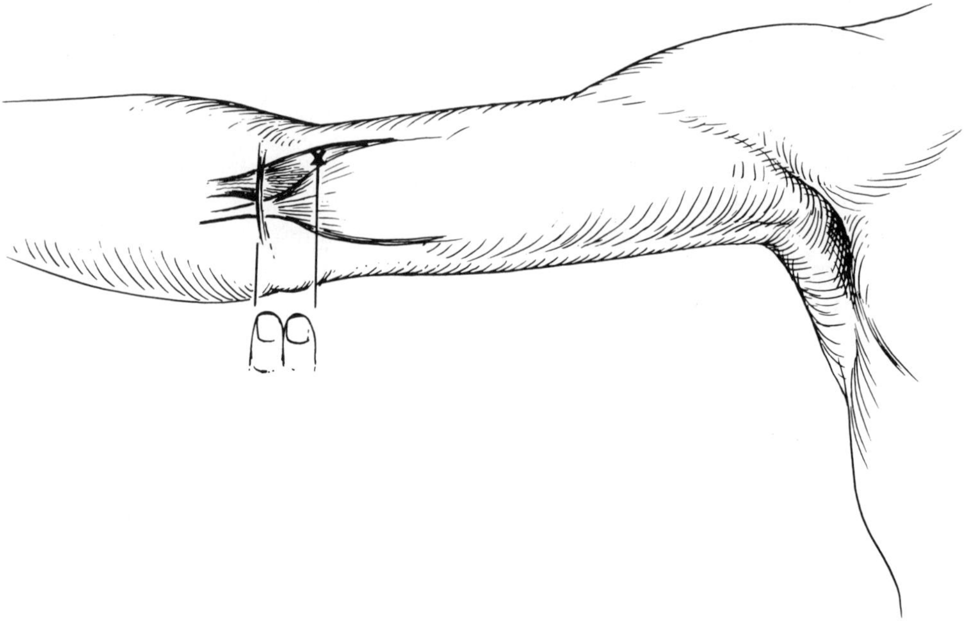

Innervation

Musculocutaneous Nerve, Lateral Cord, Anterior Division, Upper Trunk, *C5,* C6. (Also innervated by a small branch of the radial nerve.)

Origin

From the volar surface of the distal half of the humerus.

Insertion

On the tuberosity of the ulna and volar surface of the coronoid process.

Position

The patient supine with the forearm extended and pronated.

Electrode Insertion (X)

Two fingerbreadths proximal to elbow crease along and just lateral to the tendon and the bulk of the biceps.

Test Maneuver

Flex forearm with the forearm in pronation. The maneuver should be performed against minimal resistance, otherwise the biceps will be activated.

Pitfalls

If the needle electrode is inserted too medially it will be in the biceps.

Comments

(a) Most distal muscle innervated by the musculocutaneous nerve.

(b) Mainly innervated by the musculocutaneous nerve; however, receives a small amount of innervation from the radial nerve.

(c) Involved in musculocutaneous entrapment in the corocobrachialis.

CORACOBRACHIALIS

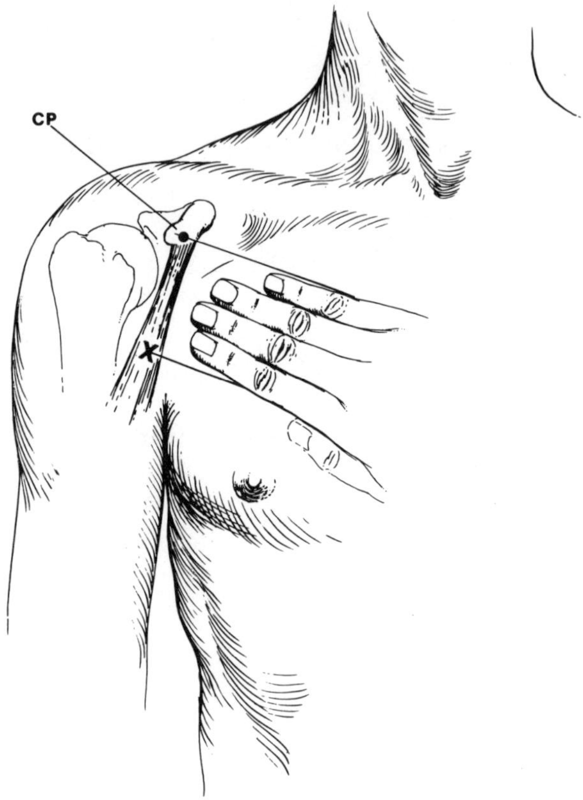

Innervation

Musculocutaneous Nerve, Lateral Cord, Anterior Division, Upper and Middle Trunk, C6, *C7*.

Origin

From the apex of the coracoid process.

Insertion

On the medial border of the humerus opposite the insertion of the deltoid.

Position

Patient supine with arm at side.

Electrode Insertion (X)

Four fingerbreadths distal to the coracoid process (CP) along volar aspect of the arm; insert needle to bone and withdraw.

Test Maneuver

With the elbow flexed to ninety degrees, forward elevation of the arm.

Pitfalls

If the needle electrode is inserted too superficially it will be in the biceps or anterior deltoid; if inserted too laterally it will be in the brachialis.

Comment

(a) The most proximal muscle innervated by the musculocutaneous nerve.

(b) The musculocutaneous nerve may be entrapped as it pierces the coracobrachialis muscle. When this occurs, this muscle is usually spared while the biceps and brachialis may be involved.

TRICEPS

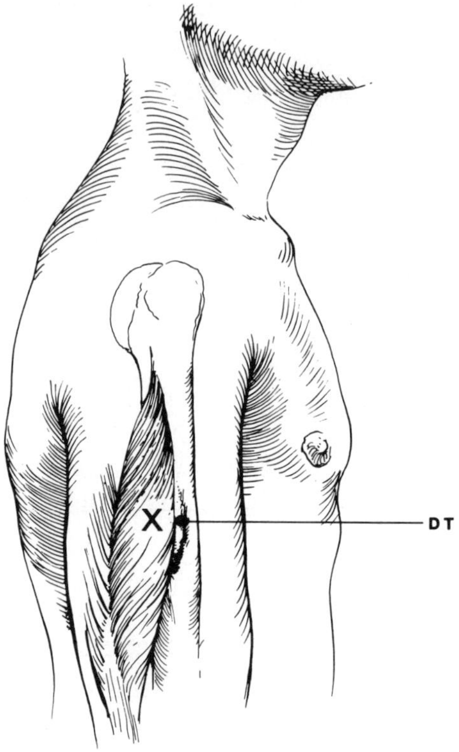

LATERAL HEAD

Innervation

Radial Nerve, Posterior Cord, Posterior Division, Middle and Lower Trunk, *C7, C8,* TI.

Origin

From the dorsal surface of the humerus above the groove for the radial nerve.

Insertion

By common tendon into distal aspect of olecranon process.

Position

Patient prone with arm abducted.

Electrode Insertion (X)

Immediately posterior to the insertion of deltoid or deltoid tubercule (DT).

Test Maneuver

Extension of elbow.

Pitfalls

If the needle electrode is inserted too anteriorly or too proximally it will be in the deltoid.

Comment

Because of its very proximal innervation through the radial nerve, it is almost never involved in "crutch paralysis" or "Saturday night palsy."

TRICEPS

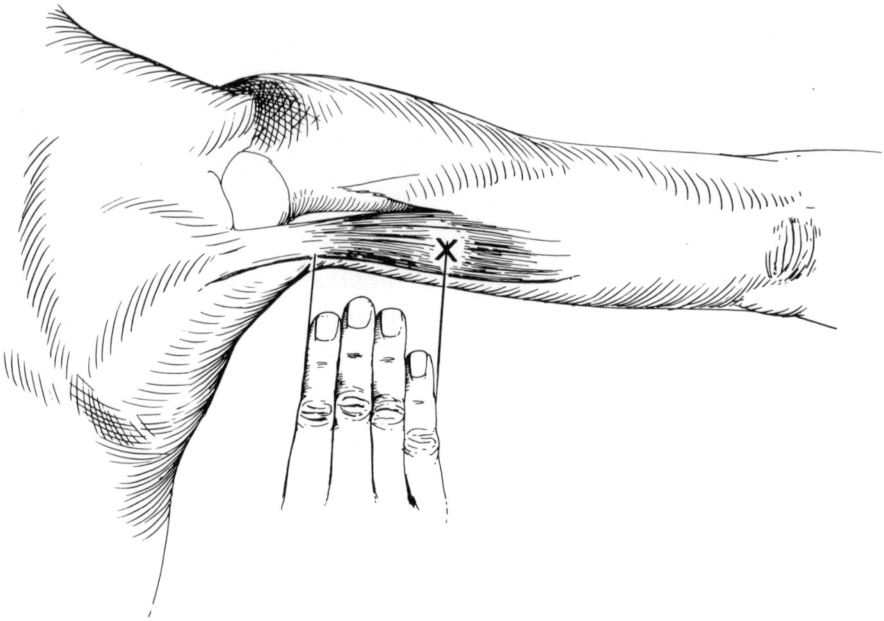

LONG HEAD

Innervation

Radial Nerve, Posterior Cord, Posterior Division, Middle and Lower Trunk, *C7, C8,* TI.

Origin

Infraglenoid tuberosity of the scapula.

Insertion

Via a common tendon, the three heads of the triceps insert on the dorsal aspect of the olecranon process of the ulna.

Position

Patient prone with arm abducted to ninety degrees and elbow flexed over edge of plinth.

Electrode Insertion (X)

Four fingerbreadths distal to the posterior axillary fold.

Test Maneuver

Extension of elbow.

Pitfalls

None.

Comment

Because of its very proximal innervation through the radial nerve, it is almost never involved in "crutch paralysis" or "Saturday night palsy."

TRICEPS

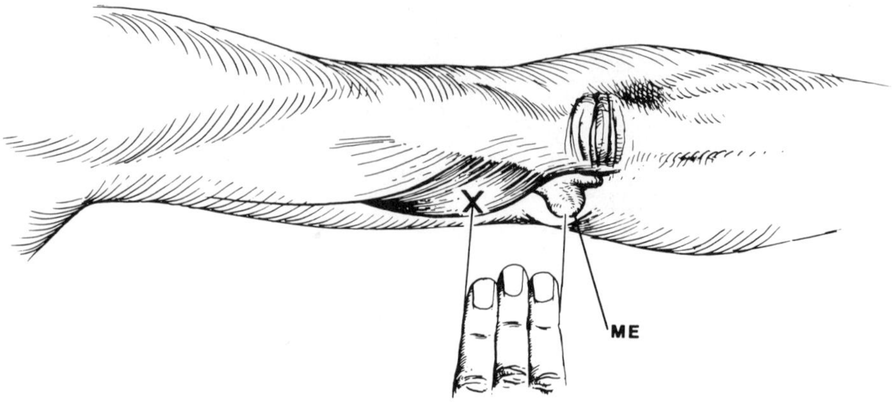

MEDIAL HEAD

Innervation

Radial Nerve, Posterior Cord, Posterior Division, Middle and Lower Trunk, *C7, C8,* TI.

Origin

From the dorsal surface of the shaft of the humerus below the groove for the radial nerve.

Insertion

By common tendon into olecranon processes.

Position

Patient prone with arm abducted.

Electrode Insertion (X)

Three fingerbreadths proximal to the medial epicondyle (ME) of humerus.

Test Maneuver

Extension of elbow.

Pitfalls

If the needle electrode is inserted too anteriorly it will be in the biceps, and there is also the danger of puncturing the brachial artery.

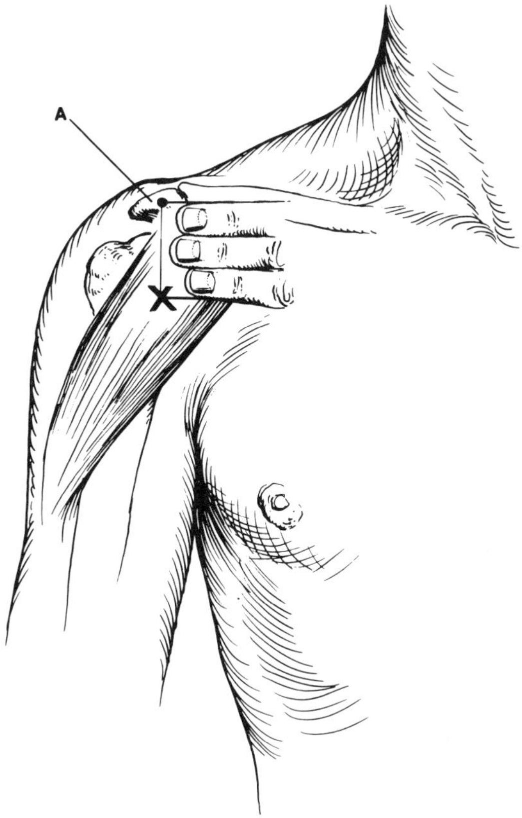

SECTION IV

SHOULDER JOINT

DELTOID, ANTERIOR

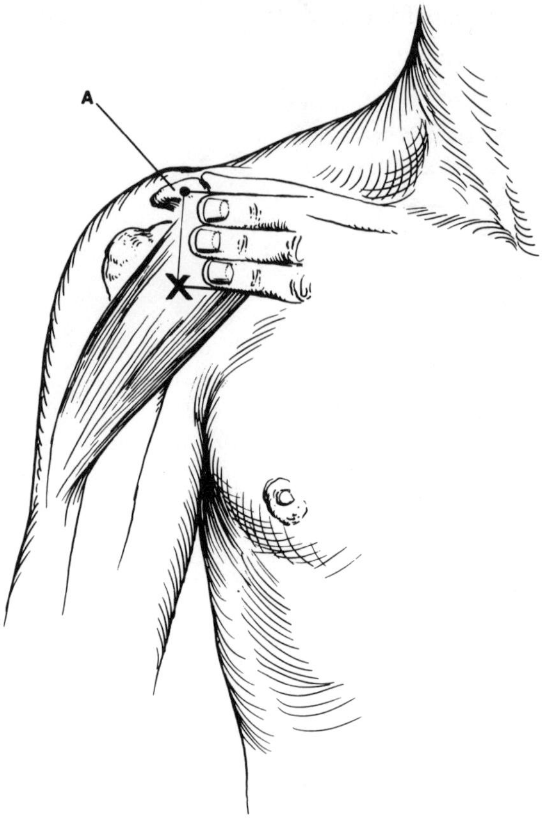

Innervation

Axillary Nerve, Posterior Cord, Posterior Division, Upper Trunk, *C5,* C6.

Origin

Lateral third of the anterior and superior surfaces of the clavicle.

Insertion

Deltoid tubercle of the humerus.

Position

Patient supine with arm at side.

Electrode Insertion (X)

Three fingerbreadths below the anterior margin of the acromion (A).

Test Maneuver

Forward elevation of the arm.

Pitfalls

If the needle electrode is inserted too medially or too deeply it will be in the coracobrachialis.

Comment

(a) Can be used as recording muscle in axillary nerve motor conduction study.
(b) If patient has a history of multiple injections into this muscle, electromyographic findings may be misleading.
(c) Involved in axillary nerve injuries secondary to fracture of surgical neck or glenohumeral joint dislocation.

DELTOID, MIDDLE

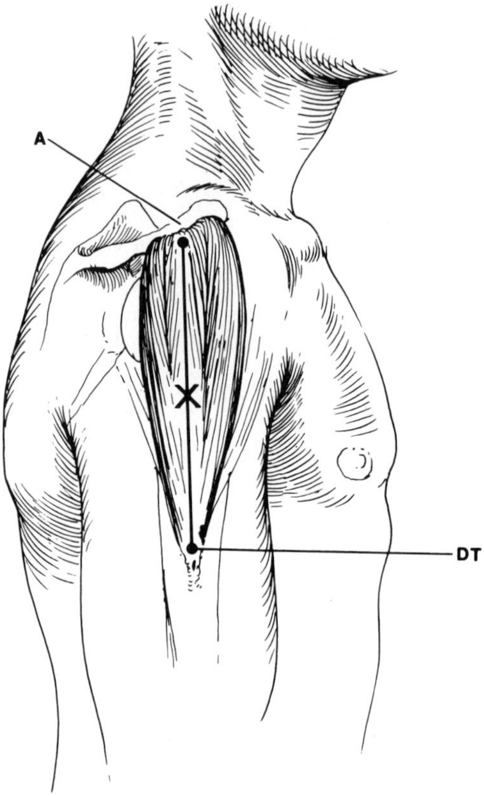

Innervation

Axillary Nerve, Posterior Cord, Posterior Division, Upper Trunk, *C5,* C6.

Origin

Acromion.

Insertion

Deltoid tubercle of the humerus.

Position

Patient supine with arm at side.

Electrode Insertion (X)

Halfway between the tip of the acromion (A) and the deltoid tubercle (DT).

Test Maneuver

Abduction of arm.

Pitfalls

None.

Comment

(a) Generally used as recording muscle in axillary nerve motor conduction study.

(b) If patient has a history of multiple injections into this muscle electromyographic findings may be misleading.

(c) Involved in axillary nerve injuries secondary to fracture of surgical neck or glenohumeral joint dislocation.

DELTOID, POSTERIOR

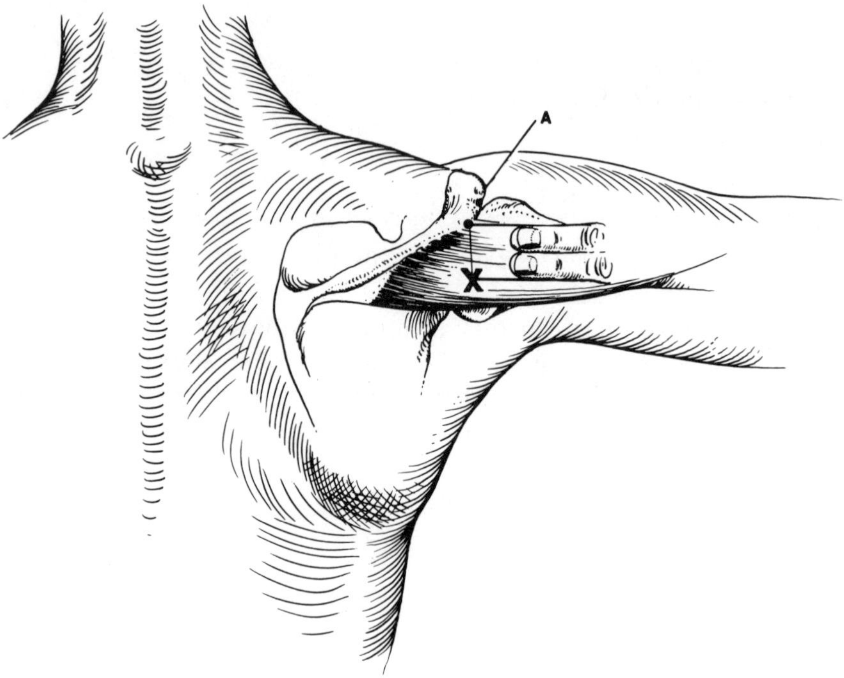

Innervation

Axillary Nerve, Posterior Cord, Posterior Division, Upper Trunk, *C5,* C6.

Origin

The spine of the scapula.

Insertion

Deltoid tubercle of the humerus.

Position

Patient prone with arm abducted to ninety degrees and elbow flexed over edge of plinth.

Electrode Insertion (X)

Two fingerbreadths caudad to posterior margin of the acromion (A).

Test Maneuver

Extension of arm.

Pitfalls

If the needle electrode is inserted too medially it will be in the teres minor; if inserted too deeply it will be in the long head of the triceps.

Comment

(a) Can be used as recording muscle in axillary nerve motor conduction study.

(b) If patient has a history of multiple injections into this muscle, electromyographic findings may be misleading.

(c) Involved in axillary nerve injuries secondary to fracture of surgical neck or glenohumeral joint dislocation.

INFRASPINATUS

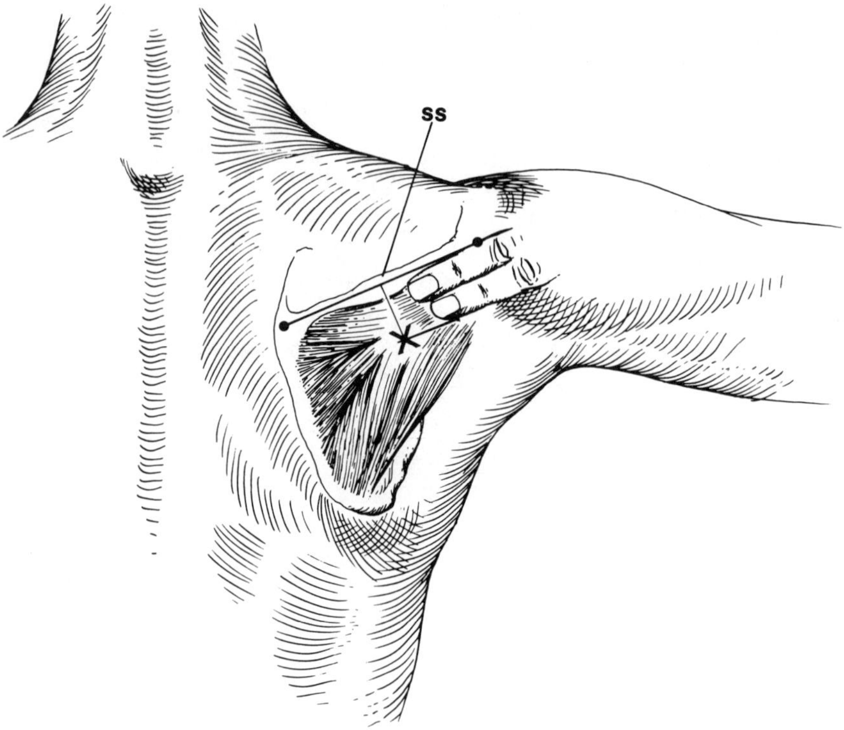

Innervation

Suprascapular Nerve, Upper Trunk, *C5,* C6.

Origin

Infraspinous fossa of scapula.

Insertion

Greater tuberosity of humerus.

Position

Patient prone with arm abducted to ninety degrees and elbow flexed over the edge of plinth.

Electrode Insertion (X)

Insert needle electrode into infraspinous fossa two fingerbreadths below medial portion of spine of scapula (SS).

Test Maneuver

Externally rotate humerus.

Pitfalls

If needle electrode is inserted too superficially it will be in the trapezius; if too laterally it will be in posterior deltoid.

Comment

(a) Used as recording muscle in suprascapular nerve conduction study.
(b) Involved in suprascapular nerve entrapment at the suprascapular notch or at the lateral edge of the spines of the scapular.
(c) Involved in Erb's type of obstetrical palsy.

LATISSIMUS DORSI

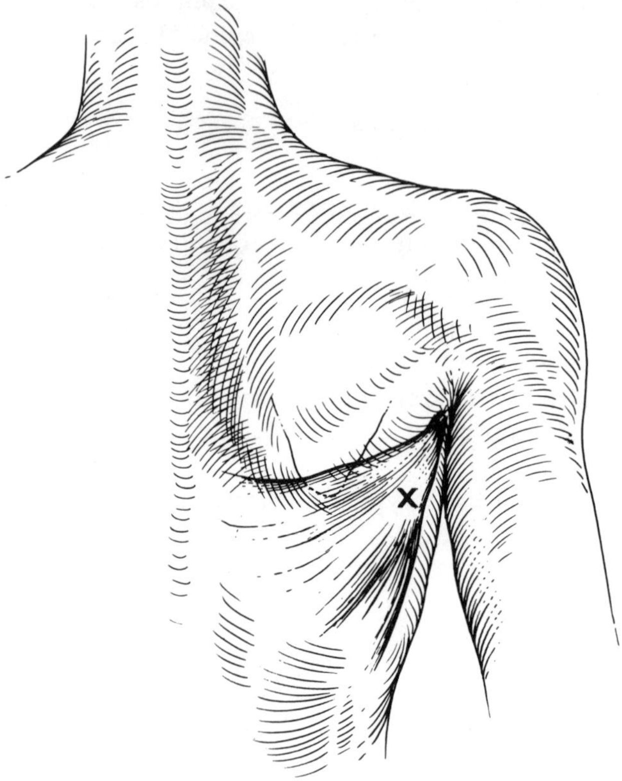

Innervation

Thoracodorsal Nerve, Posterior Cord, Posterior Division, Upper, Middle and Lower Trunk, C6, *C7*, C8.

Origin

Spinous processes of lower thoracic vertebrae, lumbodorsal fascia and posterior crest of ilium.

Insertion

Intertubercular groove of the humerus.

Position

Patient prone with arm at side and palm up.

Electrode Insertion (X)

Three fingerbreadths distal to and along posterior axillary fold.

Test Maneuver

Internally rotate, adduct and extend arm.

Pitfalls

If needle electode is inserted too medially it will be in the teres major.

PECTORALIS MAJOR

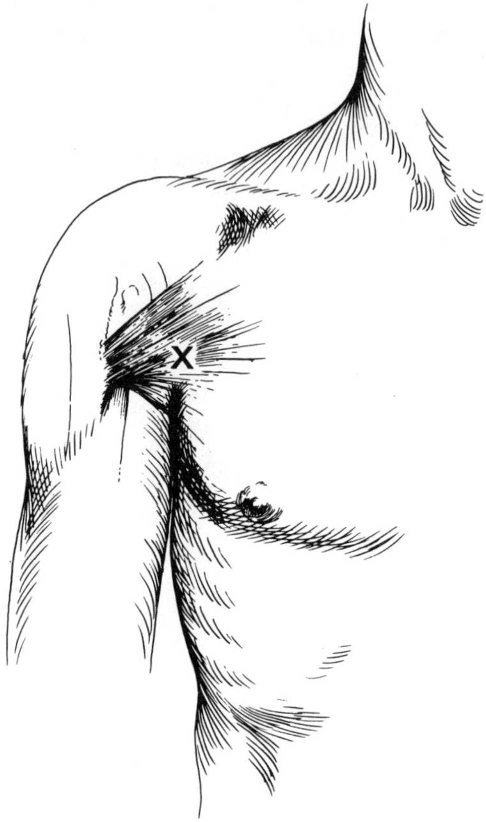

Innervation

Clavicular Portion: Lateral Pectoral Nerve, Lateral Cord, Anterior Division, Upper Trunk, C5, C6.

Sternocostal Portion: Medial Pectoral Nerve, Medial Cord, Anterior Division, Middle and Lower Trunk, C7, C8, TI.

Origin

Medial half of clavicle and anterior surface of sternum and cartilages of the first six ribs.

Insertion

The crest of the greater tubercle of humerus.

Position

Patient supine.

Electrode Insertion (X)

Insert needle electrode into anterior axillary fold.

Test Maneuver

Horizontal adduction of arm.

Pitfalls

If needle electrode is inserted too deeply it will be in the coracobrachialis; if inserted too laterally it will be in the biceps.

Comment

(a) Innervated by all segments of brachial plexus.
(b) In lateral cord lesions, the clavicular portion is involved. In medial cord lesions, sternocostal position is involved.

SUPRASPINATUS

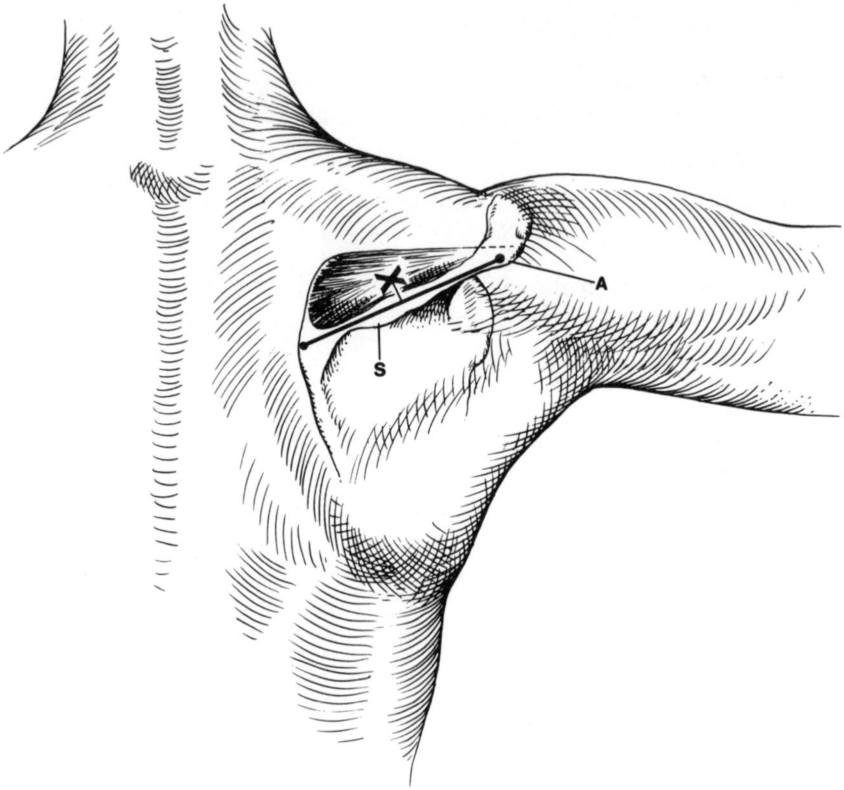

Innervation

Suprascapular Nerve, Upper Trunk, *C5,* C6.

Origins

Supraspinous fossa of scapula.

Insertion

The greater tuberosity of humerus.

Position

Patient prone with arm abducted to ninety degrees and elbow flexed over edge of plinth.

Electrode Insertion (X)

Insert into supraspinous fossa just above middle of spine of scapula (S).

Test Maneuver

Externally rotate humerus.

Pitfalls

If needle electrode is inserted too superficially it will be in the trapezius.

Comment

(a) Involved in suprascapular nerve entrapment in scapular notch.

(b) Involved in Erb's palsy.

TERES MAJOR

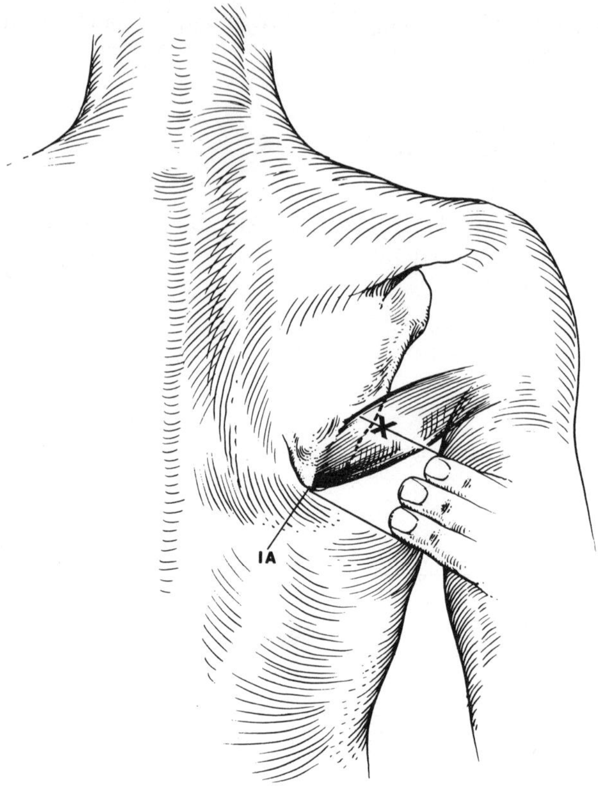

Innervation

Lower Subscapular Nerve, Posterior Cord, Posterior Division, Upper Trunk, *C5, C6.*

Origin

Inferior angle of scapula.

Insertion

The posterior bicipetal ridge.

Position

Patient prone with arm abducted to ninety degrees and elbow flexed over edge of plinth.

Electrode Insertion (X)

Three fingerbreadths above inferior angle (IA) of scapula along the lateral border.

Test Maneuver

Internally rotate humerus.

Pitfalls

If needle electrode is inserted too caudally it will be in the serratus anterior; if inserted too laterally it will be in the latissimus dorsi.

TERES MINOR

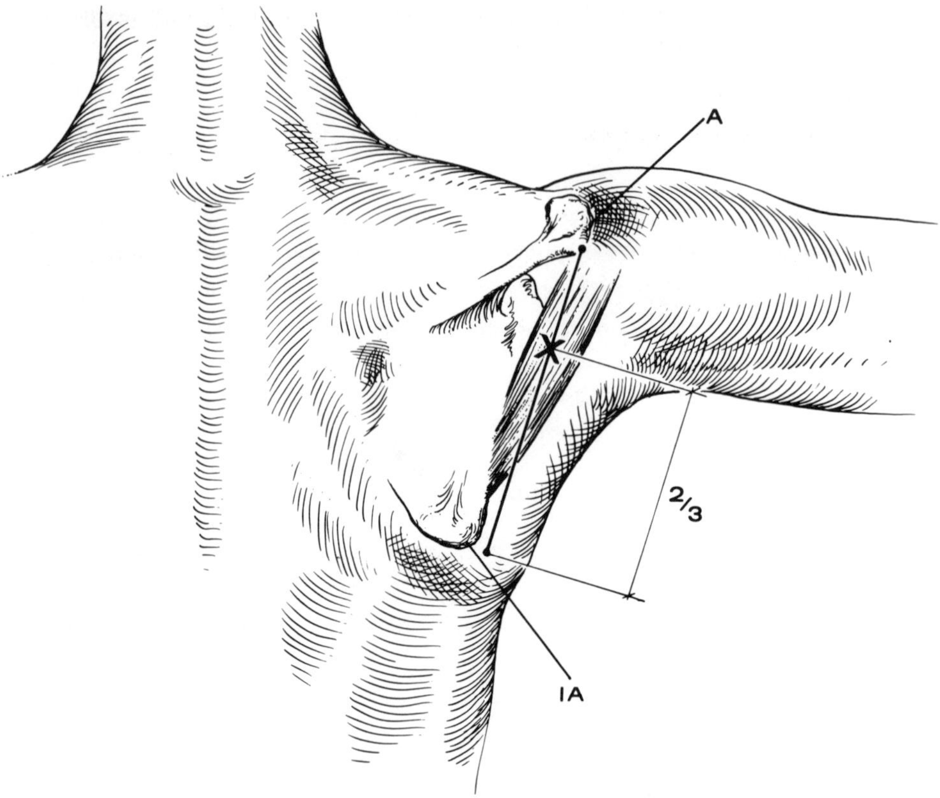

Innervation

Axillary Nerve, Posterior Cord, Posterior Division, Upper Trunk, *C5,* C6.

Origin

Upper two-thirds of axillary border of scapula.

Insertion

The greater tubercle of the humerus.

Position

Patient prone with arm abducted to ninety degrees and elbow flexed over edge of plinth.

Electrode Insertion (X)

Insert one-third of the way between acromion (A) and inferior angle (IA) of scapula along lateral border.

Test Maneuver

Externally rotate humerus.

Pitfalls

If needle electrode is inserted too cephalad it will be in the supraspinatus; if inserted too caudally it will be in the teres major or triceps; if inserted too superficially it will be in the trapezius. If needle electrode is inserted too cephalad it will be in the infraspinatus or the posterior deltoid; if inserted too caudally it will be in the teres major; if inserted too medially it will be in the infraspinatus.

Comment

(a) It is usually spared in lesions of the axillary nerve secondary to fracture of the surgical neck of the humerus or dislocation of the glenohumeral joint.
(b) Involved in lesions of axillary nerve close to posterior cord.
(c) Involved in Erb's palsy.

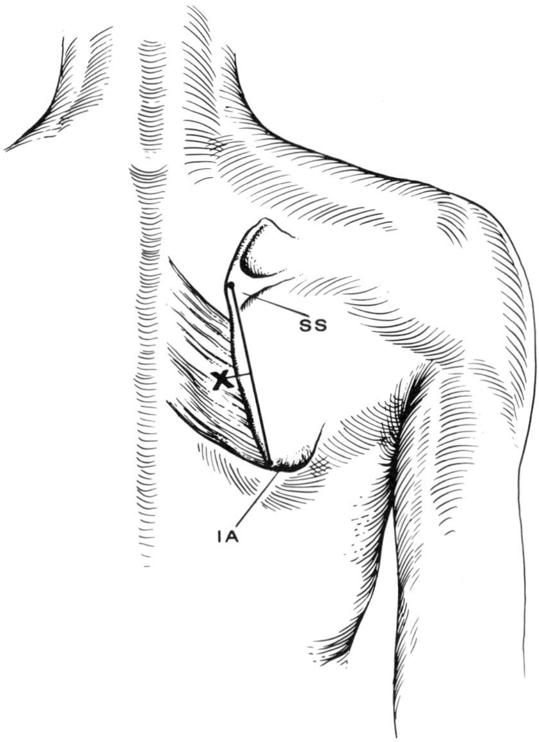

SECTION V

SHOULDER GIRDLE

LEVATOR SCAPULAE

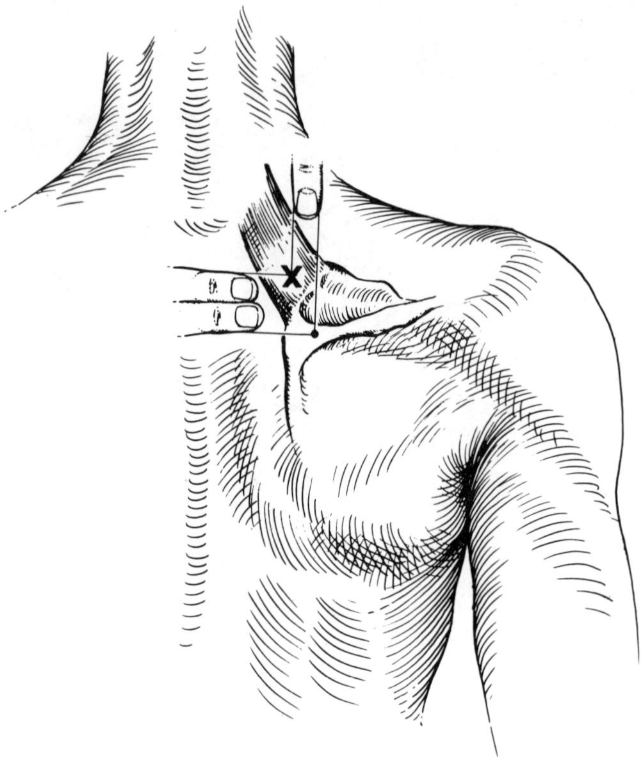

Innervation

Dorsal Scapular Nerve (C5) Plus Twigs From C3, C4.

Origin

Transverse processes of upper four cervical vertebrae.

Insertion

Vertebral border of scapular above root of spine.

Position

Patient prone.

Electrode Insertion (X)

Two fingerbreadths cephalad to the medial angle of scapula and one fingerbreadth medial.

Test Maneuver

Elevate scapulae.

Pitfalls

If needle electrode is inserted too superficially it will be in the trapezius; if inserted too deeply it will be in the paraspinal muscles.

Comment

This muscle is innervated directly and mainly from C5 root and therefore is an ideal muscle to test integrity of the C5 root.

PECTORALIS MINOR

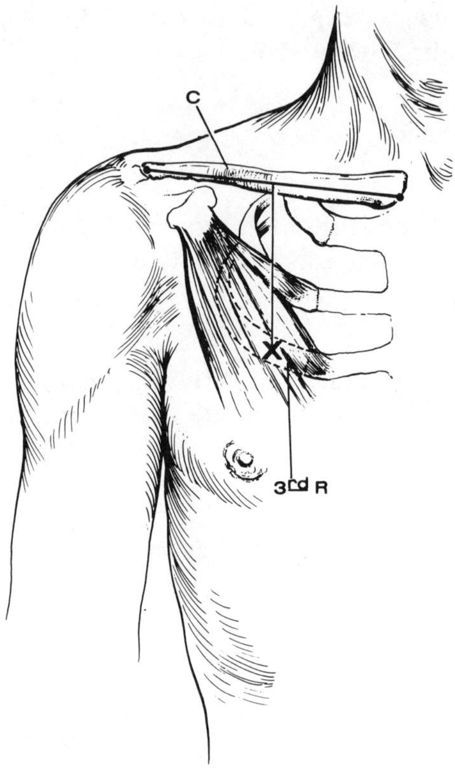

Innervation

Medial and Lateral Pectoral Nerve, Medial and Lateral Cords, Anterior Division, Upper, Middle and Lower Trunks, *C6, C7, C8.*

Origin

Anterior surface of the third to the fifth ribs.

Insertion

Upon the coracoid process of the scapula.

Position

Patient supine.

Electrode Insertion (X)

In midclavicular line insert needle electrode to anterior surface of the third rib (3rd R) and withdraw slightly.

Test Maneuver

Scapula depression.

Pitfalls

If needle electrode is inserted too superficially it will be in the pectoralis major.

RHOMBOIDEUS MAJOR

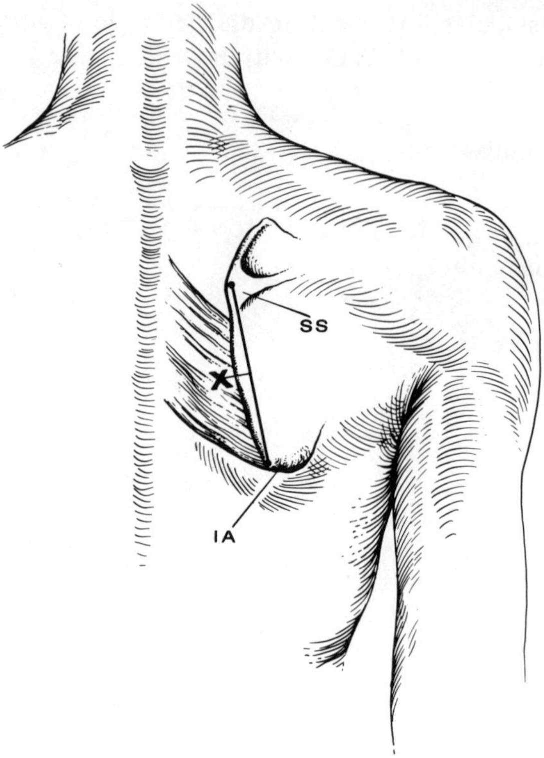

Innervation

Dorsal Scapular Nerve, *C5.*

Origin

Spinous processes of the second to fifth thoracic vertebrae.

Insertion

Vertebral border of scapula: from root of spine of scapula to its inferior angle.

Position

Patient prone with arm internally rotated so that dorsum of hand rests on small of back.

Electrode Insertion (X)

Midway between spine (SS) and inferior angle (IA) of scapula just medial to vertebral border.

Test Maneuver

Raise hand from small of back.

Pitfalls

If needle electrode is inserted too superficially it will be in the trapezius; if inserted too deeply it will be in the erector spinae muscles.

Comment

Along with the rhomboideus minor, innervated directly and solely from the C5 nerve root; therefore ideal muscles to test for the integrity of the C5 root.

RHOMBOIDEUS MINOR

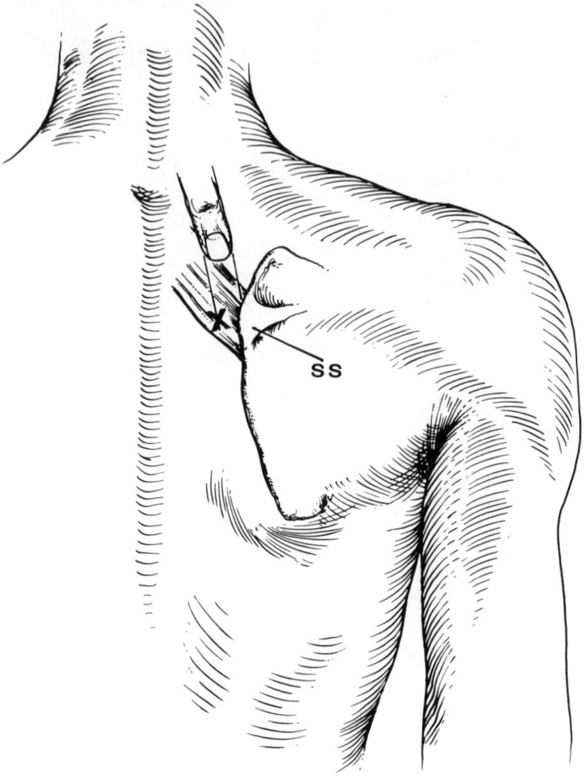

Innervation

Dorsal Scapular Nerve, *C5.*

Origin

Spinous processes of seventh cervical and first thoracic vertebrae.

Insertion

Vertebral border of scapula at the base of the spine of the scapula.

Position

Patient prone with arm internally rotated so that dorsum of hand rests on small of back.

Electrode Insertion (X)

One fingerbreadth medial to vertebral end of scapular spine (SS).

Test Maneuver

Raise hand from small of back.

Pitfalls

If needle electrode is inserted too deeply it will be in the serratus posterior superior.

Comment

See rhomboiseus major.

SERRATUS ANTERIOR

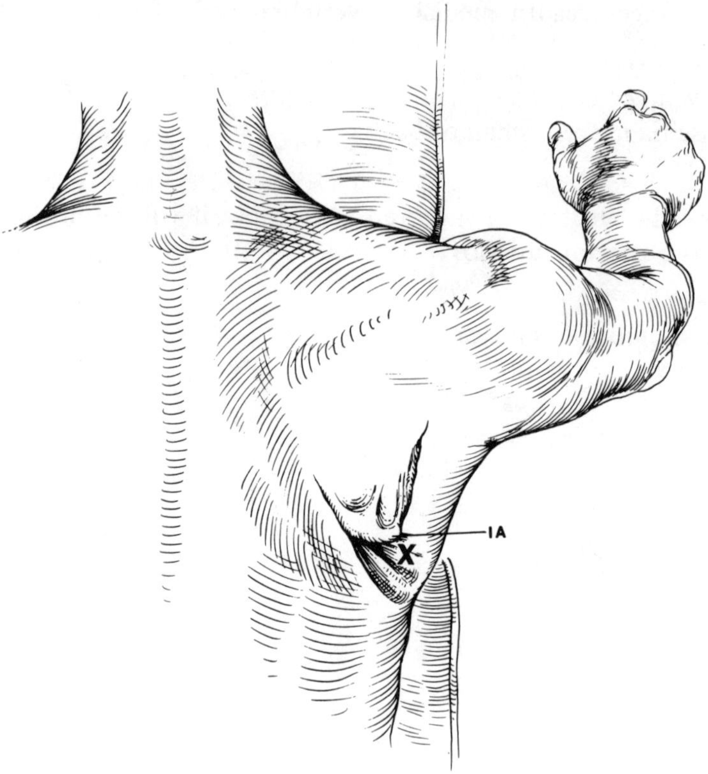

Innervation

Long Thoracic Nerve, *C5, C6, C7.*

Origin

Digitations from anterior surfaces and superior borders of upper nine ribs.

Insertion

Ventral surface of vertebral border of scapula, from superior to inferior angles.

Position

Patient prone with arm dangling over edge of plinth.

Electrode Insertion (X)

Just lateral to inferior angle (IA) of scapula.

Test Maneuver

Patient presses hand against resistance.

Pitfalls

If needle electrode is inserted too superficially it will be in the latissimus dorsi; if inserted too cephalad it will be in the teres major.

Comment

Innervated directly from roots C5, C6, C7 through long thoracic nerve; therefore a good muscle to study when trying to distinguish root from more distal lesions.

TRAPEZIUS, LOWER

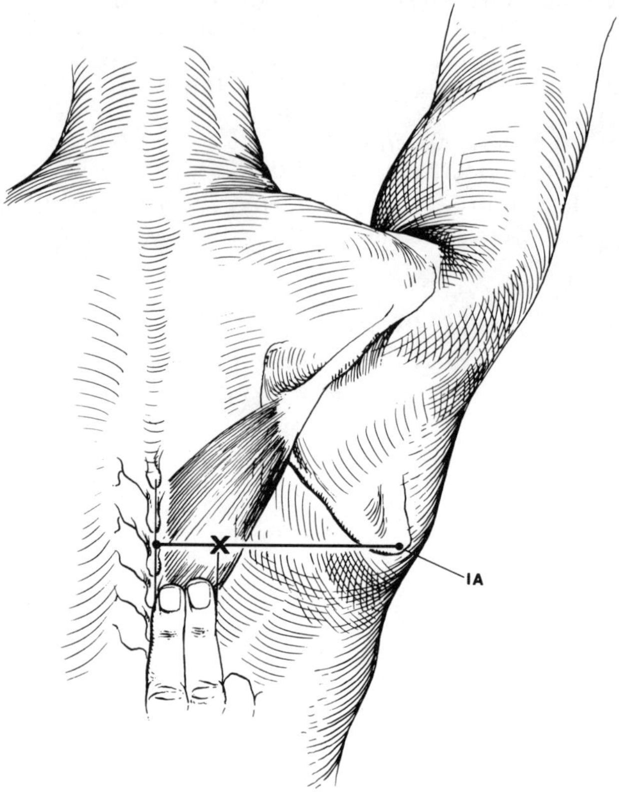

Innervation

Spinal Portion of Accessory Nerve and Twigs from C3, C4.

Origin

Spinous processes of lower thoracic vertebrae.

Insertion

The spine of the scapula.

Position

Patient prone with arm extended overhead.

Electrode Insertion (X)

On a line perpendicular to the vertebral column at the level of the inferior angle (IA) of the scapula, two fingerbreadths from the spinous process of that vertebra.

Test Maneuver

Elevate arm from plinth.

Pitfalls

If needle electrode is inserted too deeply or too caudally it will be in the latissimus dorsi.

Comment

This muscle may be involved due to injury to its innervation in cervical lymph node biopsy.

TRAPEZIUS, MIDDLE

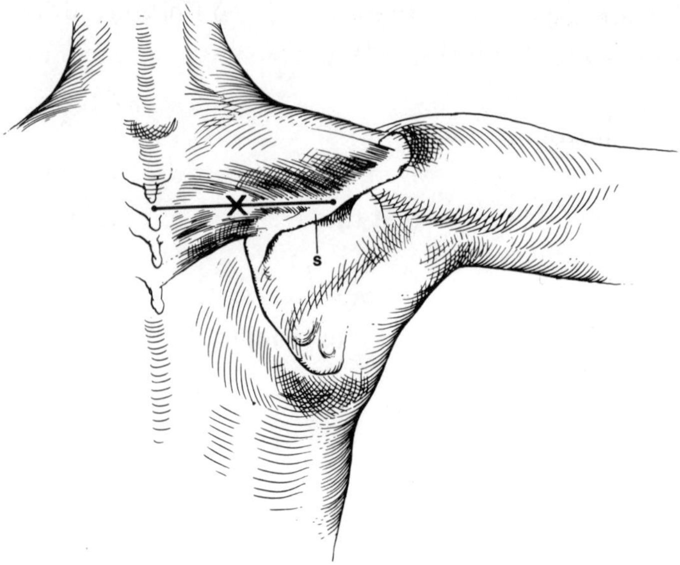

Innervation

Spinal Portion of Accessory Nerve and Twigs From C3 and C4.

Origin

The seventh cervical and spinous processes of upper thoracic vertebrae.

Insertion

The acromion process and spine of scapula.

Position

The patient prone with arm abducted to ninety degrees and elbow flexed over the edge of plinth.

Electrode Insertion (X)

Midway between the midpoint of spine (S) of scapula and spinous process of vertebra at the same level.

Test Maneuver

Adduct scapula by elevation of arm from plinth.

Pitfalls

If needle electrode is inserted too deeply it will be in the rhomboideus.

Comment

This muscle may be involved due to injury to its innervation in cervical lymph node biopsy.

TRAPEZIUS, UPPER

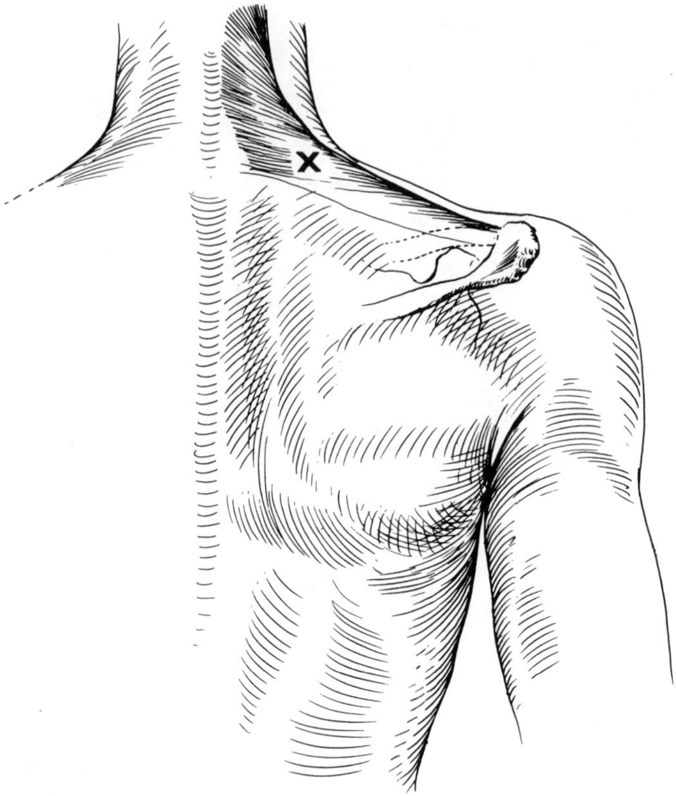

Innervation

Spinal Portion of Accessory Nerve and Twigs from C3 and C4.

Origin

Occipital bone and ligamentum nuchae.

Insertion

The outer third of clavicle.

Position

The patient prone.

Electrode Insertion (X)

At angle of neck and shoulder.

Test Maneuver

Shrug shoulder.

Pitfalls

If needle electrode is inserted too deeply it will be in the levator scapula.

Comment

This muscle may be involved due to injury to its innervation in cervical lymph node biopsy.

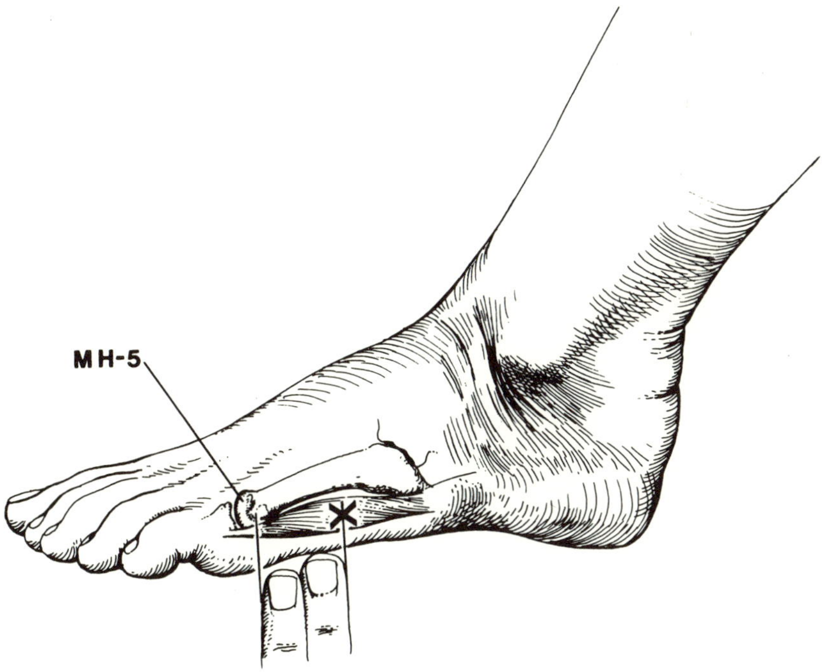

SECTION VI

FOOT

ABDUCTOR DIGITI QUINTI

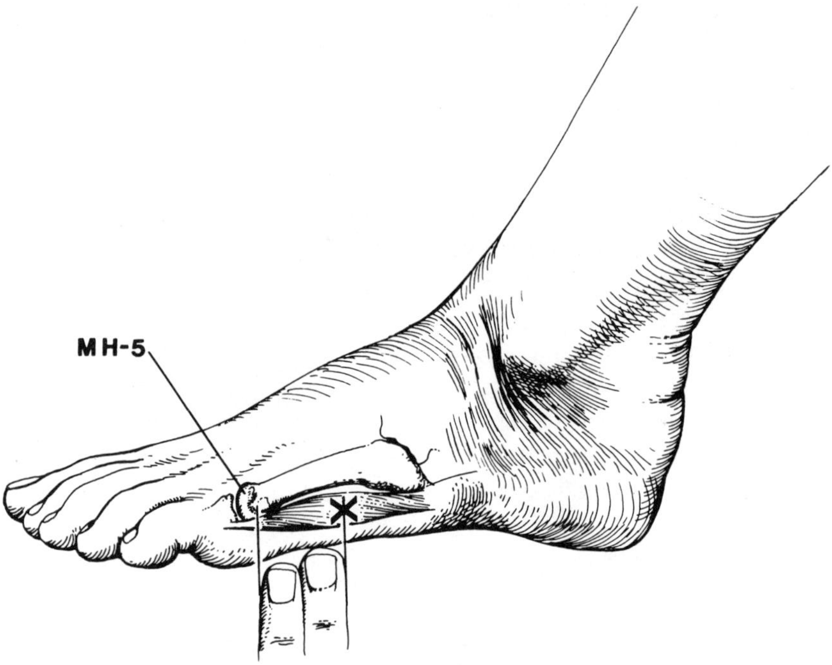

Innervation

Lateral Plantar Nerve, Tibial Nerve, Sciatic Nerve, Ventral Division Sacral Plexus, S1, S2.

Origin

From the lateral process of the tuberosity of the calcaneus.

Insertion

To the lateral side of the base of the proximal phalanx of the fifth toe.

Position

The patient supine.

Electrode Insertion (X)

On the lateral border of the foot, two fingerbreadths proximal to the head of the fifth metatarsal (MH-5).

Test Maneuver

Patient to spread the toes.

Pitfalls

None.

Comment

(a) Commonly involved in –

1. Peripheral neuropathy secondary to diabetes mellitus
2. Lateral plantar nerve lesion
3. Tarsal-tunnel syndrome idiopathic or following ankle fractures
4. More proximal lesions involving the tibial or sciatic nerve, sacral plexus or S1, S2 roots.

(b) This muscle can be used as recording muscle for the tibial nerve conduction studies; the distal latency will indicate the functional status of the lateral plantar nerve.

ABDUCTOR HALLUCIS

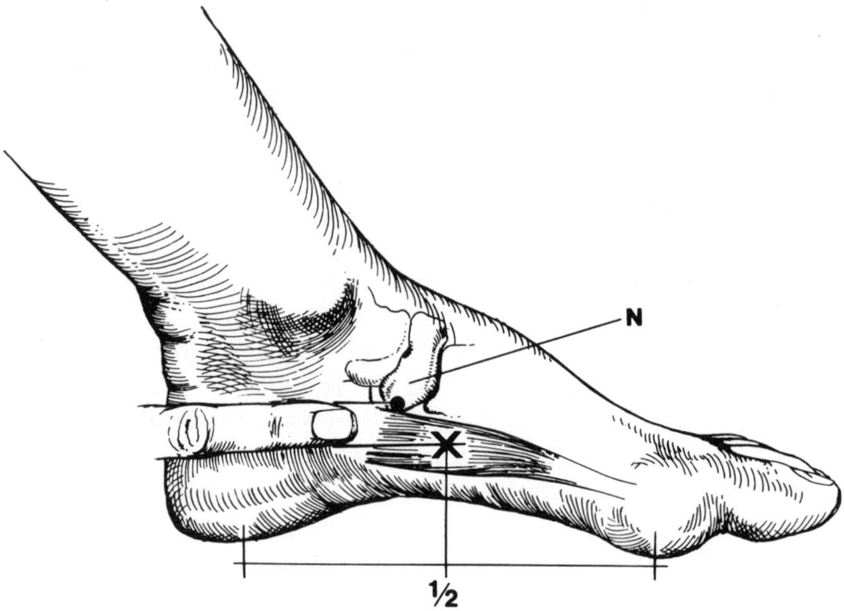

Innervation

Medial Plantar Nerve, Tibial Nerve, Sciatic Nerve, Ventral Division Sacral Plexus, S1–S2.

Origin

From the medial tuberosity of the calcaneus.

Insertion

To the medial side of the base of the proximal phalanx of the great toe.

Position

The patient supine.

Electrode Insertion (X)

One fingerbreadth below the navicular (N) bone on midportion of medial aspect of foot.

Test Maneuver

Patient to spread the toes.

Pitfalls

If the electrode is inserted too distally it will be in the flexor pollicis brevis; if inserted too deeply it will be in the flexor digitorum brevis.

Comment

(a) Commonly involved in –

1. Peripheral neuropathy secondary to diabetes mellitus
2. Medial plantar nerve lesions
3. Tarsal tunnel syndrome, idiopathic or post ankle fracture
4. More proximal lesions involving the tibial or sciatic nerve, sacral plexus or S1, S2 roots.

(b) This muscle can be used as recording muscle for the tibial nerve conduction studies; the distal latency will indicate the functional status of the medial plantar nerve.

ADDUCTOR HALLUCIS

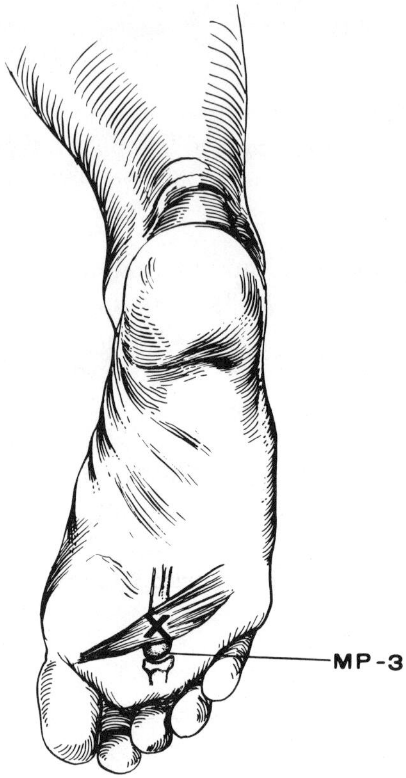

Innervation

Lateral Plantar Nerve, Tibial Nerve, Sciatic Nerve, Ventral Division Sacral Plexus, S1, S2.

Origin

From the peroneus longus tendon sheath and adjacent parts of the cuboid and second, third and fourth metatarsal bones (oblique head), and from the capsule of the third, fourth and fifth metatarsalphalangeal joints (transverse head).

Insertion

Into the lateral aspect of the base of the proximal phalanx of the big toe.

Position

The patient supine.

Electrode Insertion (X)

For the transverse head only: Insert the electrode just proximal to the third metatarsalphalangeal (MP-3) joint until the metatarsal bone is felt; withdraw the electrode slightly.

Test Maneuver

Passively abduct big toe and ask patient to adduct.

Pitfalls

If the electrode is inserted too superficially it will be in the lumbrical muscle.

Comments

Involved in peripheral neuropathy secondary to diabetes mellitus.

EXTENSOR DIGITORUM BREVIS

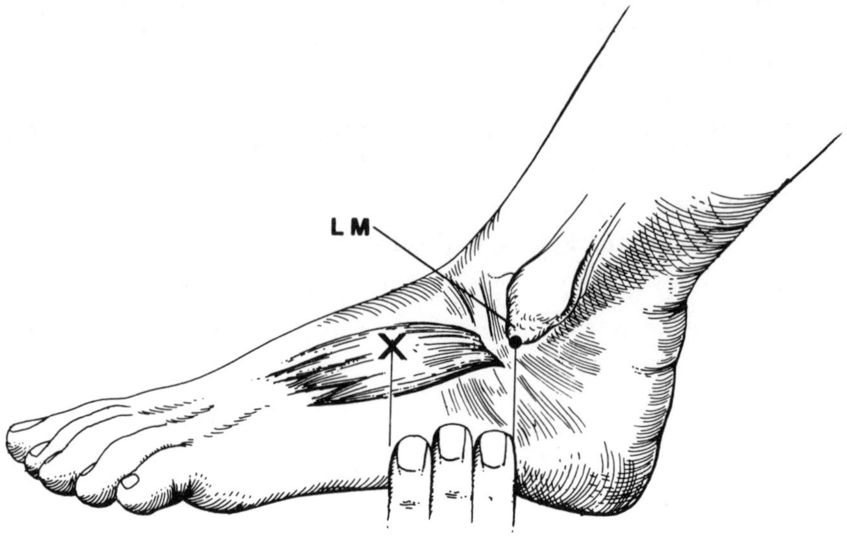

Innervation

Deep Peroneal Nerve, Common Peroneal Nerve, Sciatic Nerve, Dorsal Division Sacral Plexus, L5, S1.

Origin

From the upper and lateral surface of the calcaneus.

Insertion

Into the base of the proximal phalanx of the great toe and the tendon of the extensor digitorum longus of the second, third and fourth toes.

Electrode Insertion (X)

Three fingerbreadths distal to the lower border of the lateral malleolus (LM) parallel to the lateral border of the foot.

Test Maneuver

Patient to extend the toes.

Pitfalls

None.

Comment

(a) Last muscle innervated by common peroneal nerve.

(b) This muscle is used as recording muscle for the common peroneal nerve conduction study.

(c) Involved in—

1. Lesions of peripheral neuropathy secondary to diabetes mellitus
2. Deep peroneal nerve lesions
3. More proximal nerve lesions of common peroneal nerve, sciatic nerve, sacral plexus, or L5, S1 nerve roots.

FLEXOR DIGITORUM BREVIS

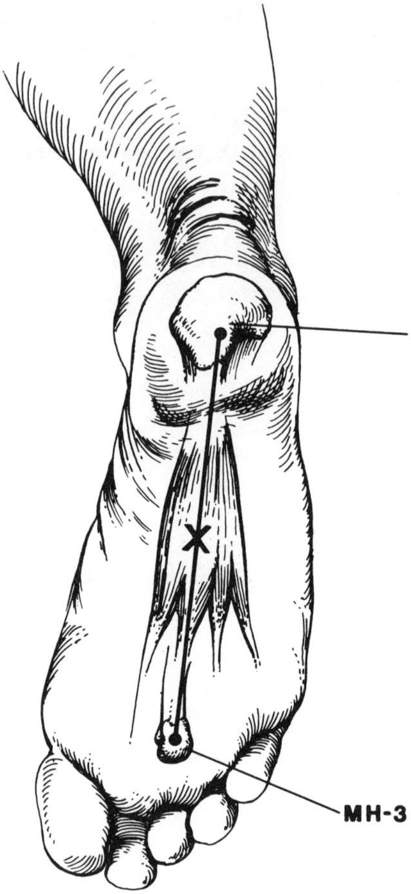

Innervation

Medial Plantar Nerve, Tibial Nerve, Sciatic Nerve, Ventral Division Sacral Plexus, S1, S2.

Origin

From the medial tubercle of the calcaneus and the deep surface of the plantar aponeurosis.

Insertion

Into the plantar surface of the base of the middle phalanx of the second, third, fourth and fifth toes.

Position

Patient supine.

Electrode Insertion (X)

The electrode is inserted midway between the third metatarsal head (MH-3) and the calcaneous (C) to the plantar aponeurosis and then withdrawn slightly.

Test Maneuver

Patient to flex the toes.

Pitfalls

If the electrode is inserted too laterally it will be in the abductor digiti minimi; if inserted too medially it will be in the abductor hallucis brevis; if inserted too deeply it will be in the quadratus plantae.

Comment

Commonly involved in—

(1) Peripheral neuropathy secondary to diabetes mellitus

(2) Medial plantar nerve lesions

(3) Tarsal tunnel syndrome

(4) More proximal lesions involving the tibial nerve, sciatic nerve, sacral plexus or S1, S2 roots.

FLEXOR DIGITI MINIMI

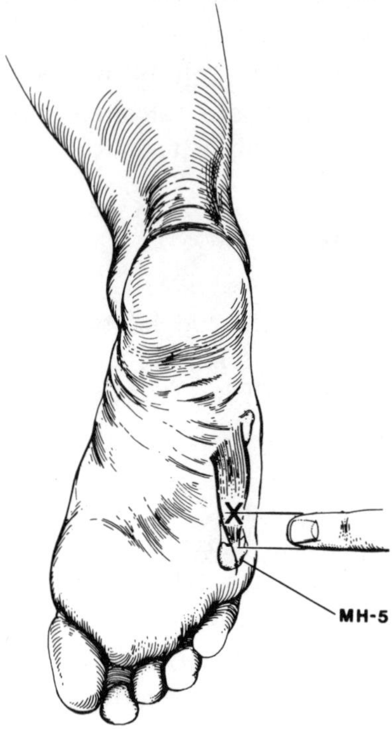

Innervation

Lateral Plantar Nerve, Tibial Nerve, Sciatic Nerve, Ventral Division Sacral Plexus, S1, S2.

Origin

From the sheath of the peroneus longus tendon and the base of the fifth metatarsal bone.

Insertion

In a common tendon with the abductor digit minimi into the lateral side of the base of the proximal phalanx of the little toe.

Position

The patient supine.

Electrode Insertion (X)

On the plantar surface of the foot, one fingerbreadth proximal to the fifth metatarsal head (MH-5), the electrode is inserted to the bone and then withdrawn slightly.

Test Maneuver

Patient to flex the metatarsophalangeal joint of the little toe.

Pitfalls

If the electrode is inserted too laterally it will be in the abductor digiti minimi; if inserted too medially it will be in the lumbricals.

Comments

Commonly involved in—

(1) Peripheral neuropathy secondary to diabetes mellitus
(2) Lateral plantar nerve
(3) Tarsal tunnel syndrome
(4) More proximal lesions involving the tibial nerve, sciatic nerve, sacral plexus or S1, S2 roots.

FLEXOR HALLUCIS BREVIS

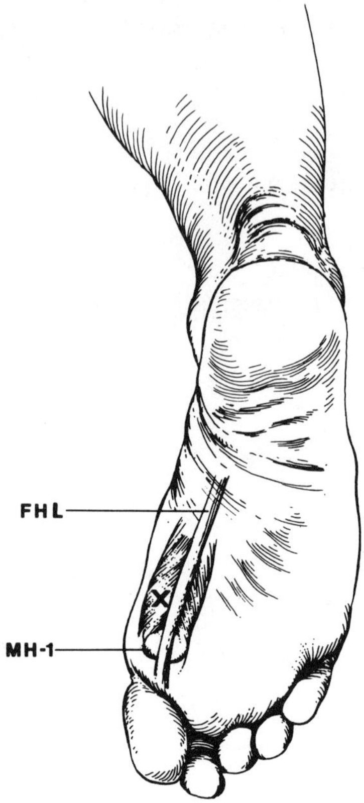

Innervation

Medial Plantar Nerve, Tibial Nerve, Sciatic Nerve, Ventral Division Sacral Plexus, S1, S2.

Origin

The two muscle bellies originate from a tendinous expansion of the tibialis posterior muscle insertion and from the cuneiform and the cuboid.

Insertion

The two tendons end at each side of the base of the proximal phalanx of the big toe, the medial one in a common tendon with that of the abductor, and the lateral one in a common tendon with that of the adductor hallucis brevis.

Position

The patient supine.

Electrode Insertion (X)

For medial head only: Proximal and medial to the tendon of the flexor hallucis longus.

Test Maneuver

Patient to flex the metatarsalphalangeal joint of the big toe.

Pitfalls

If the electrode is inserted too laterally it will be in the adductor hallucis; if inserted too medially it will be in the abductor hallucis.

Comment

Commonly involved in—

(1) Peripheral neuropathy secondary to diabetes mellitus
(2) Medial plantar nerve
(3) Tarsal-tunnel syndrome
(4) More proximal lesions involving the tibial nerve, sciatic nerve, sacral plexus or S1, S2 roots.

INTEROSSEI[3]

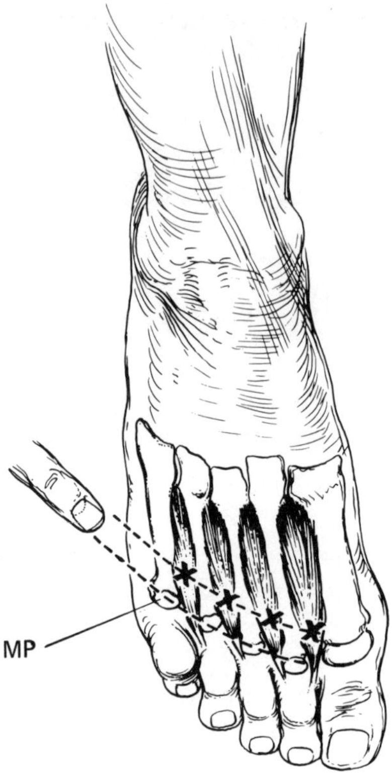

Innervation

Lateral Plantar Nerve, Tibial Nerve, Sciatic Nerve, Sacral Plexus, S1, S2.

Origin

From the shaft of the metatarsal into the base of the proximal phalanges and the dorsal digital expansions.

Position

The patient supine.

[3] It was impossible to differentiate between the dorsal and volar interossei with the needle electrode.

Electrode Insertion (X)

The needle electrode is inserted one fingerbreadth proximal to the metatarsalphalangeal (MP) joints in the intermetatarsal space. The number of interosseus muscle corresponds to the number of the intermetatarsal space.

Test Maneuver

Spread toes.

Pitfalls

If the needle electrode is inserted too deeply it will be either in the oblique or transverse head of the adductor hallucis.

Comment

This group of muscles is commonly involved in—

(1) Peripheral neuropathy secondary to diabetes mellitus
(2) Lateral plantar nerve lesions
(3) Tarsal-tunnel syndrome
(4) More proximal lesions involving the tibial nerve, sciatic nerve, sacral plexus, or S1, S2 roots.

QUADRATUS PLANTAE

(FLEXOR DIGITORUM ACCESSORIUS)

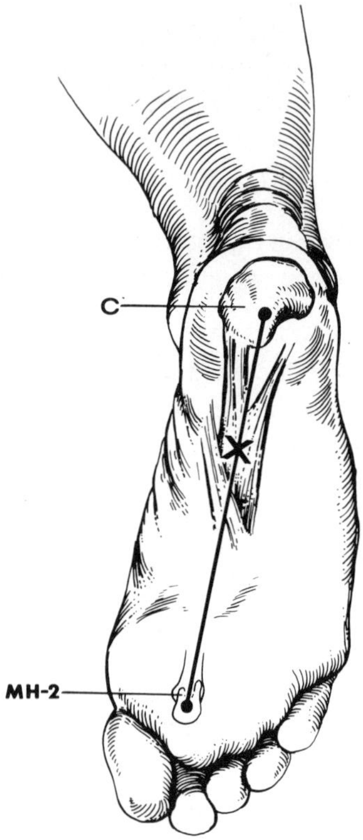

Innervation

Lateral Plantar Nerve, Tibial Nerve, Sciatic Nerve, Ventral Division Sacral Plexus, S1, S2.

Origin

Medial Head: From medial surface of calcaneous.

Lateral Head: From lateral border of the plantar surface of the calcaneous.

Insertion

Into the tendon of the flexor digitorum longus.

Position

The patient supine.

Electrode Insertion (X)

Insert the electrode at the junction of proximal and middle one-third of a line between the tip of the calcaneous (C) and the second metatarsal head (MH-2). Insert deep to bone and withdraw slightly.

Test Maneuver

Patient to flex the toes.

Pitfalls

If the electrode is inserted too superficially it will be in the flexor digitorum brevis; if inserted too laterally it will be in the abductor digiti minimi; if inserted too medially it will be in the abductor hallucis brevis.

Comment

Involved in lesions of–

1. Peripheral neuropathy secondary to diabetes mellitus
2. Lateral plantar nerve lesions
3. Tarsal-tunnel syndrome
4. More proximal lesions of tibial nerve, sciatic nerve, sacral plexus and S1, S2 nerve roots.

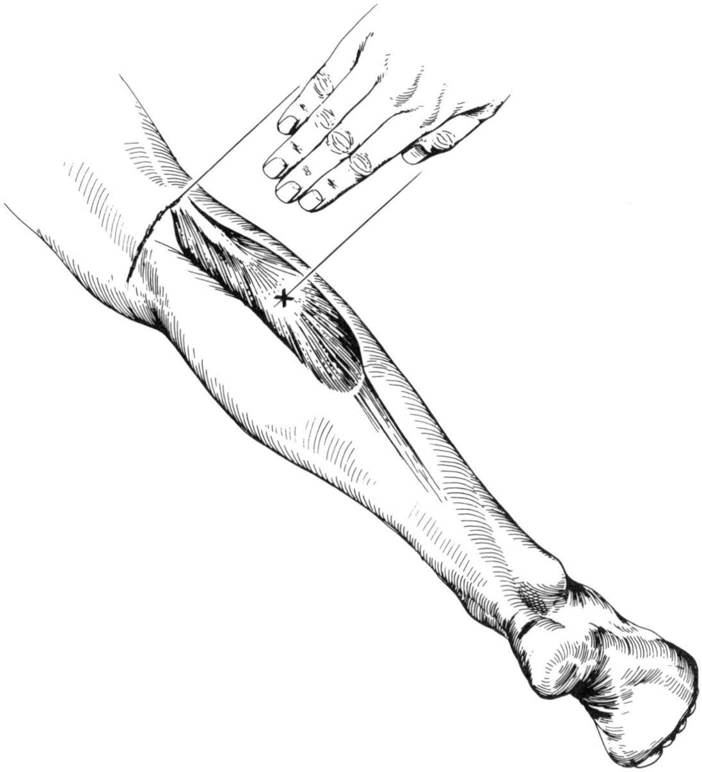

SECTION VII
LEG

EXTENSOR DIGITORUM LONGUS

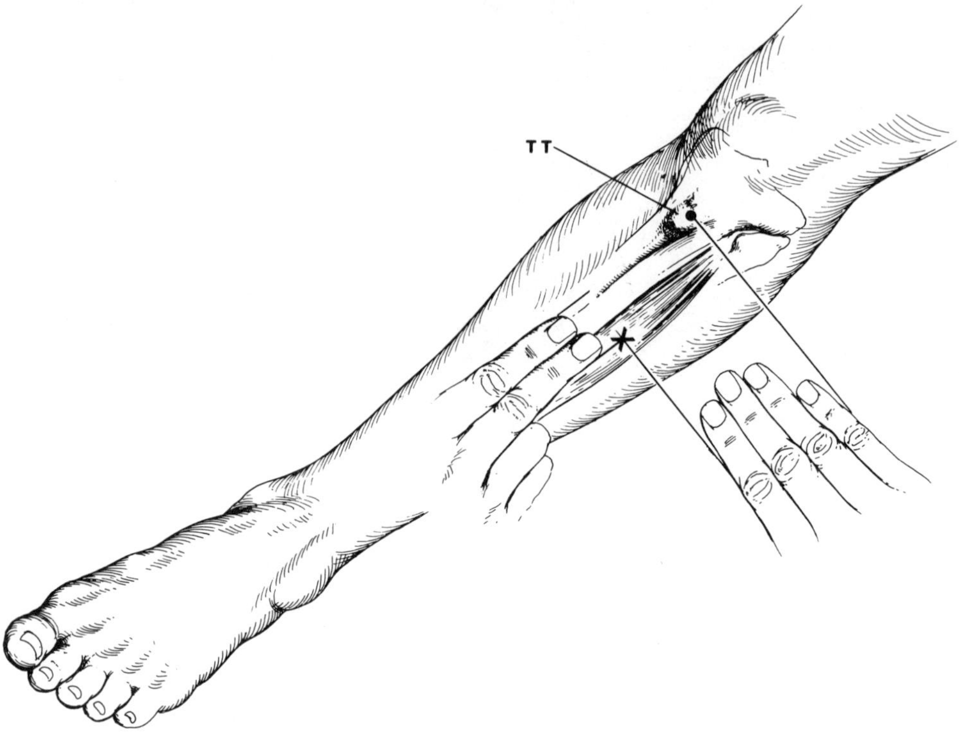

Innervation

Deep Peroneal Nerve, Common Peroneal Nerve, Sciatic Nerve, Posterior Division Sacral Plexus, L5, S1.

Origin

From the lateral condyle of the tibia, the proximal three-fourths of the fibula and the interosseus membrane.

Insertion

Through a common tendon with the lumbricals and the interossei into the dorsum of the middle and distal phalanx of the four lateral toes.

Position

The patient supine.

Electrode Insertion (X)

Four fingerbreadths distal to the tibial tubercle (TT) and two fingerbreadths lateral to the tibial crest, the electrode is inserted through the tibialis anterior about one inch.

Test Maneuver

Patient to extend the four lateral toes.

Pitfalls

It is a very difficult muscle to examine because it is covered by the tibialis anterior and the peroneal muscles. If the electrode is inserted too anteriorly it will be in the tibialis anterior; if inserted too laterally it will be in the peroneus longus, and if inserted too superficially it will be in the tibialis anterior.

Comment

Involved in–

1. Anterior compartment syndrome
2. Lesions of the deep peroneal nerve
3. Common peroneal nerve
4. Sciatic nerve
5. Sacral plexus.

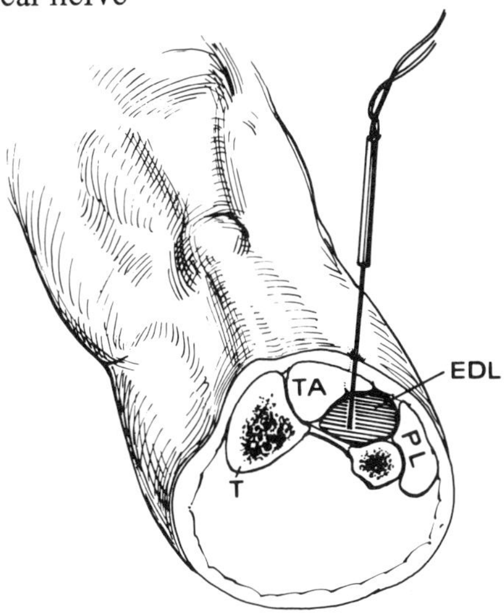

EDL – Extensor Digitorum Longus
TA – Tibialis Anterior
PL – Peroneus Longus
T – Tibia

EXTENSOR HALLUCIS LONGUS

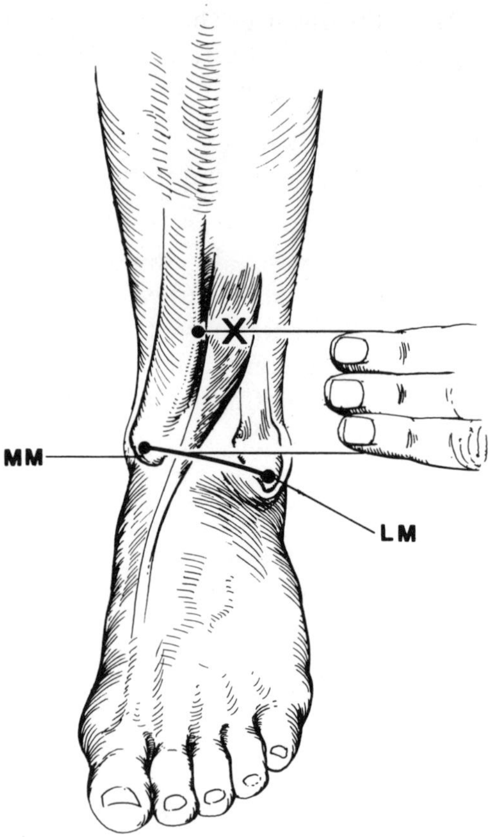

Innervation

Deep Peroneal Nerve, Common Peroneal Nerve, Sciatic Nerve, Posterior Division Sacral Plexus, L5, *S1*.

Origin

From the midportion of the shaft of the fibula.

Insertion

Into the distal phalanx of the great toe.

Position

The patient supine.

Electrode Insertion (X)

Three fingerbreadths above the bimalleolar line (MM-LM) of the ankle just lateral to the crest of the tibia.

Test Maneuver

Patient to extend the big toe.

Pitfalls

If the electrode is inserted too superficially and too proximally it will be in the tibialis anterior; if inserted too laterally it will be in the peroneus tertius.

Comment

Involved in–

1. Anterior compartment syndrome
2. Lesion of the deep peroneal nerve
3. Common peroneal nerve
4. Sciatic nerve
5. Sacral plexus
6. L5, S1 root lesions.

FLEXOR DIGITORUM LONGUS

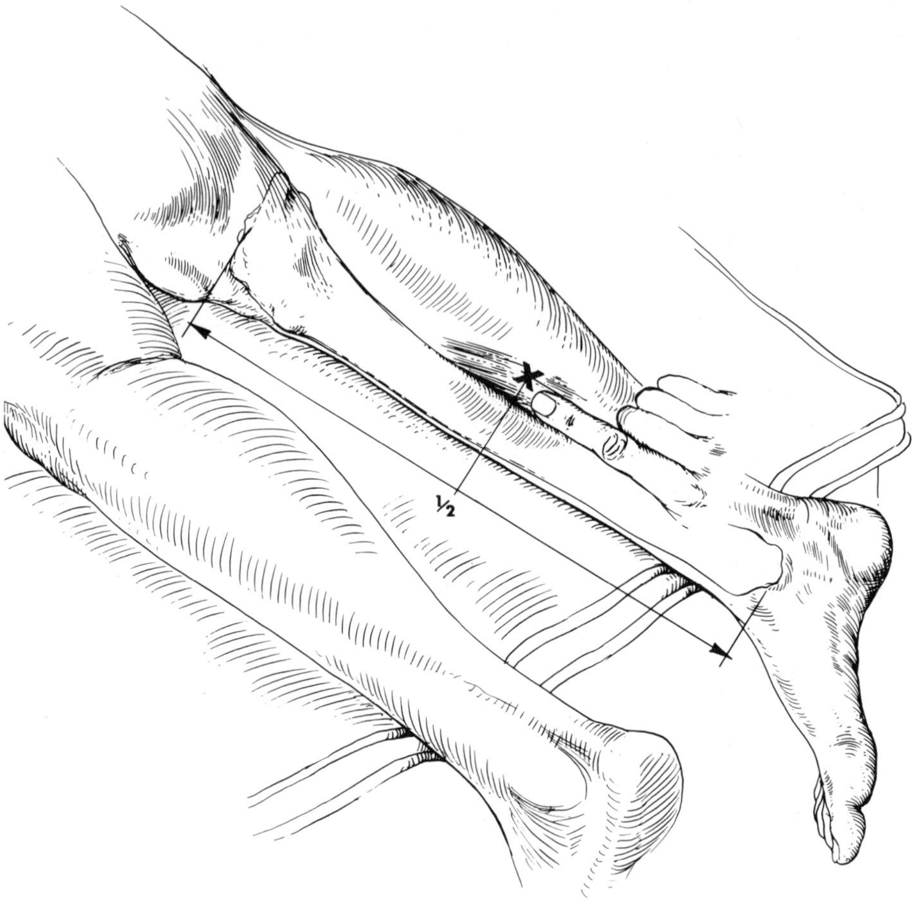

Innervation

Tibial Nerve, Sciatic Nerve, Ventral Division Sacral Plexus, *L5*, *S1*, S2.

Origin

From the body of the tibia below the popliteal line.

Insertion

Into the base of the distal phalanges of the second, third, fourth and fifth toe.

Position

The patient prone.

Electrode Insertion (X)

Palpate the medial edge of tibia at midshaft and insert electrode just posterior to it.

Test Maneuver

Patient to flex the toes, without flexing the ankle.

Pitfalls

If the electrode is inserted too superficially it will be in the soleus; if inserted too deeply it will be in the tibialis posterior.

Comment

Involved in lesions of:

1. The tibial nerve
2. Sciatic nerve
3. Sacral plexus
4. L5, S1, S2 roots.

FLEXOR HALLUCIS LONGUS

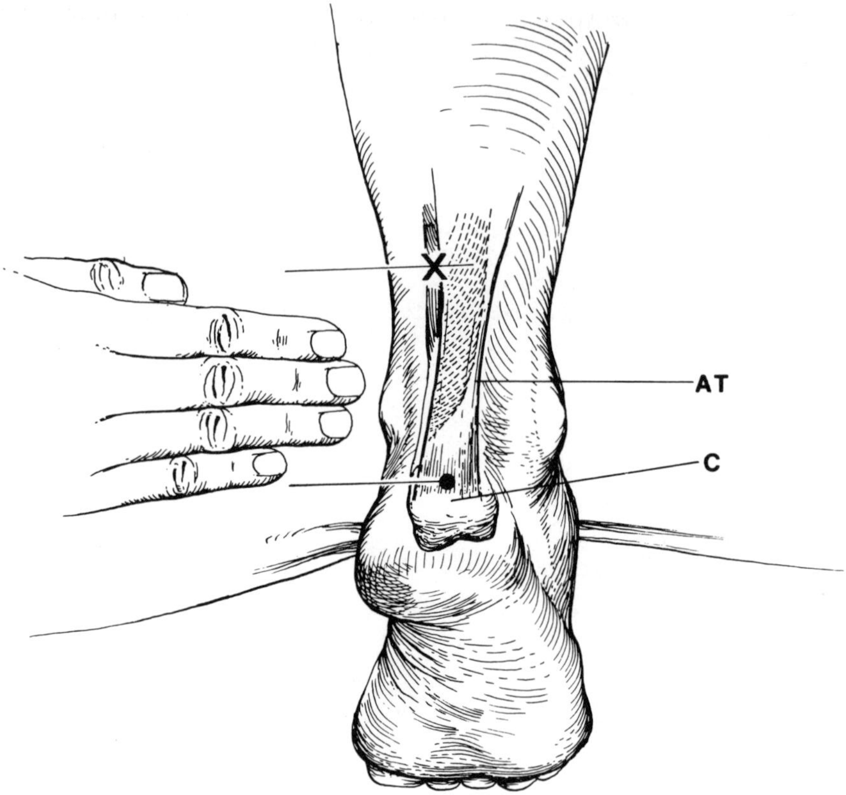

Innervation

Tibial Nerve, Sciatic Nerve, Ventral Division Sacral Plexus, L5, S1, S2.

Origin

From the inferior two-thirds of the posterior surface of body of the fibula and the interosseus membrane.

Insertion

Into the base of the distal phalanx of the great toe.

Position

The patient prone.

Electrode Insertion (X)

Insert the electrode obliquely five fingerbreadths above the insertion of the Achilles tendon (AT) and anterior to the medial border of this tendon towards the tibia.

Test Maneuver

Patient to flex the big toe, keeping the ankle and the small toes relaxed.

Pitfalls

If the electrode is inserted too deeply it will be in the tibialis posterior; if inserted too anteriorly it will be in the flexor digitorum longus; if inserted too superficially it will be in the soleus.

Comment

Involved in lesions of:

1. Tibial nerve
2. Sciatic nerve
3. Sacral plexus
4. L5, S1, S2 roots.

GASTROCNEMIUS: Lateral Head

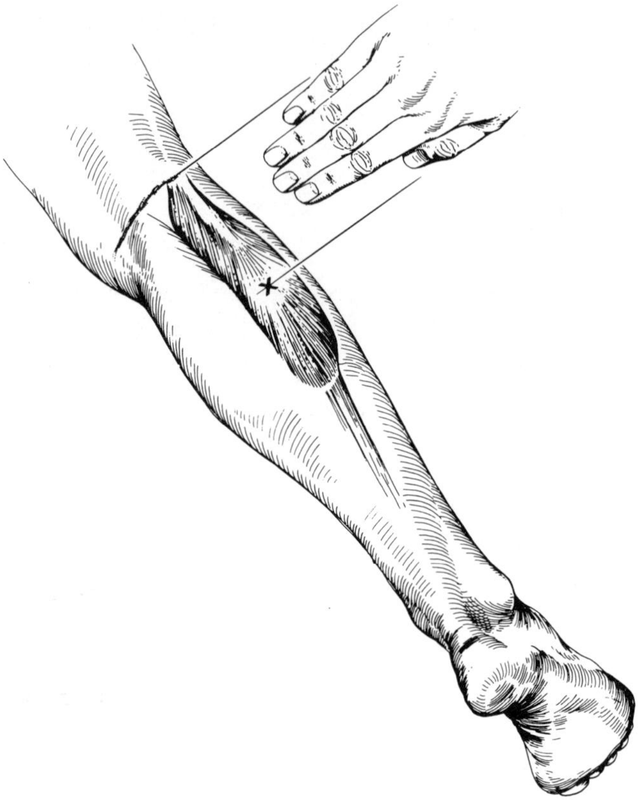

Innervation

Tibial Nerve, Sciatic Nerve, Ventral Division Sacral Plexus, S1, S2.

Origin

From the lateral femoral condyle.

Insertion

Into the calcaneous, through the Achille's tendon.

Position

The patient prone with feet over edge of plinth.

Electrode Insertion (X)

One handbreadth below the popliteal crease on the lateral mass of the calf.

Test Maneuver

Patient to plantar flex the foot with the knee extended.

Pitfalls

If the electrode is inserted too deeply it will be in the soleus.

Comment

Involved in lesions of:

1. Tibial nerve
2. Sciatic nerve
3. Sacral plexus
4. L5, S1, S2 roots.

GASTROCNEMIUS: Medial Head

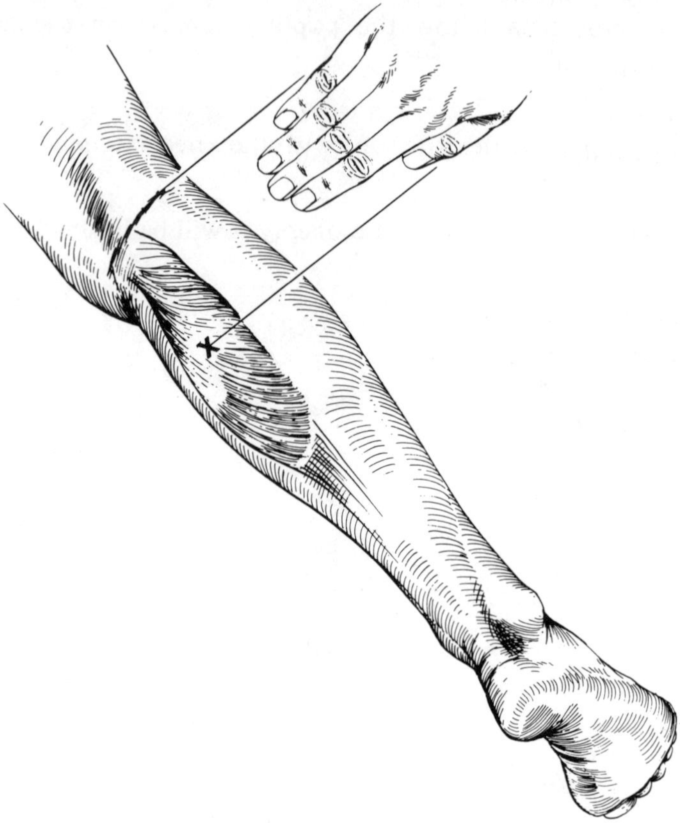

Innervation

Tibial Nerve, Sciatic Nerve, Ventral Division Sacral Plexus, S1, S2.

Origin

From the medial femoral condyle.

Insertion

Into the calcaneous, through the Achille's tendon.

Position

The patient prone with feet over edge of plinth.

Electrode Insertion (X)

One hand breadth below the popliteal crease on the medial mass of the calf.

Test Maneuver

Patient to plantar flex the foot with the knee extended.

Pitfalls

If the electrode is inserted too deeply it will be in the flexor digitorum longus.

Comment

Involved in lesions of:

1. Tibial nerve
2. Sciatic nerve
3. Sacral plexus
4. L5, S1, S2 roots.

PERONEUS BREVIS

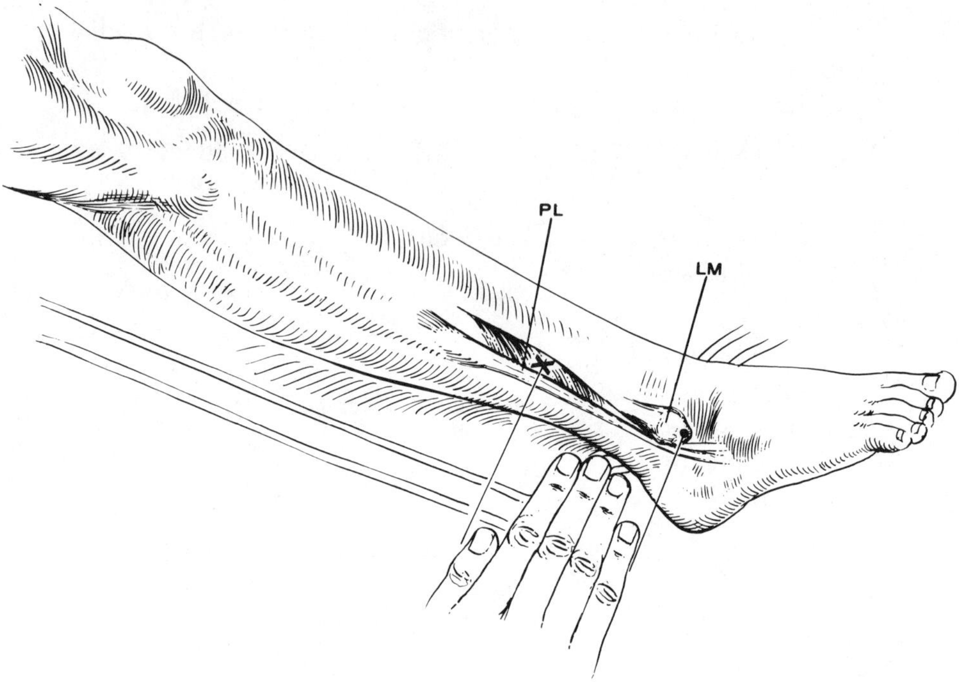

Innervation

Superficial Peroneal Nerve, Sciatic Nerve, Posterior Division Sacral Plexus, *L5, S1,* S2.

Origin

From the lower two-thirds of the fibula.

Insertion

Into the base of the fifth metatarsal head.

Position

The patient supine.

Electrode Insertion (X)

One handbreadth proximal to the lateral malleolus (LM) and anterior to the peroneus longus (PL) tendon.

Test Maneuver

Patient to plantar flex and evert the foot.

Pitfalls

If the electrode is inserted too proximally it will be in the peroneus longus; if inserted too anteriorly it will be either in the peroneus tertius or in the extensor digitorum longus.

Comment

(a) Can be used as recording muscle when the superficial peroneal nerve is to be studied.

(b) Involved in lesions of:

1. Superficial peroneal nerve
2. Common peroneal nerve
3. Sciatic nerve
4. Sacral plexus
5. L5, S1 roots.

PERONEUS LONGUS

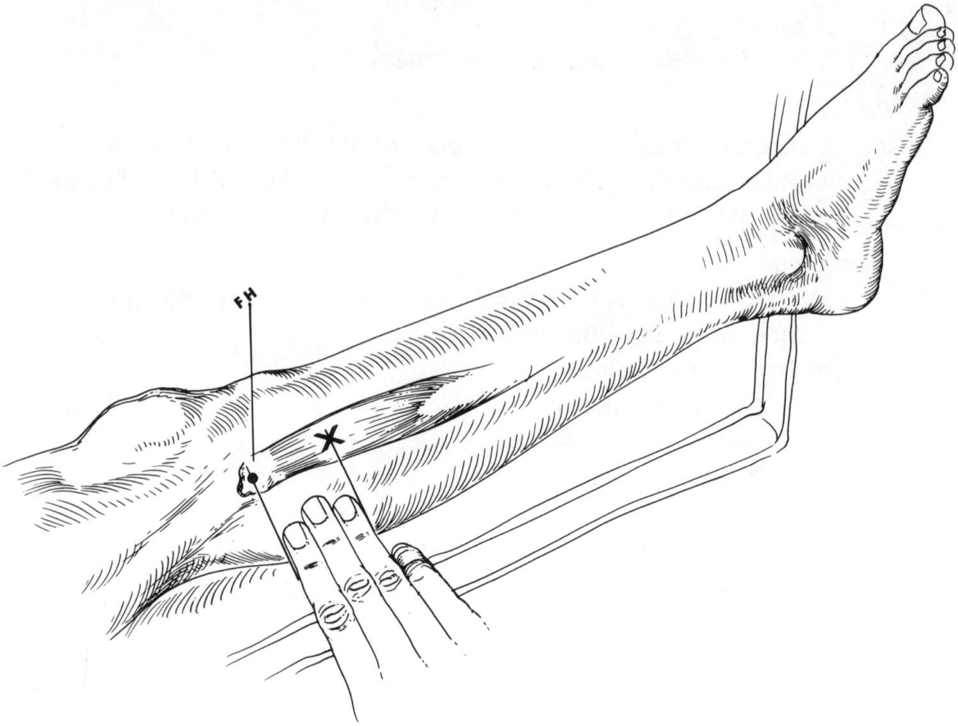

Innervation

Superficial Peroneal Nerve, Sciatic Nerve, Posterior Division Sacral Plexus, *L5, S1,* S2.

Origin

From the fibular head and from the proximal two-thirds of the fibula.

Insertion

Into the base of the first metatarsal and the first cuneiform.

Position

The patient supine.

Electrode Insertion (X)

Three fingerbreadths below the fibular head (FH) directed toward the lateral aspect of the fibula.

Test Maneuver

Patient to plantar flex and evert the foot.

Pitfalls

If the electrode is inserted too posteriorly it will be in the soleus; if inserted too anteriorly it will be in the extensor digitorum longus.

Comment

This muscle is involved in lesions of:

1. Superficial peroneal nerve
2. Common peroneal nerve
3. Sciatic nerve
4. Sacral plexus
5. L5, S1 roots.

PERONEUS TERTIUS

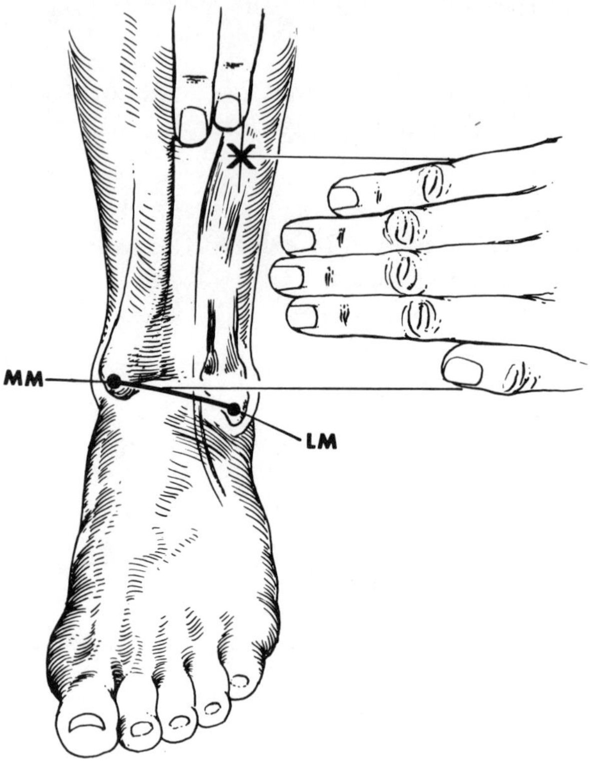

Innervation

Deep Peroneal Nerve, Common Peroneal Nerve, Sciatic Nerve, Posterior Division Sacral Plexus, L5, S1.

Origin

From the lower one-third of the fibula.

Insertion

Into the base of the fifth metatarsal head.

Position

The patient supine.

Electrode Insertion (X)

One handbreadth above bimalleolar line (MM-LM) of the ankle and two fingerbreadths lateral to the tibia.

Test Maneuver

Patient to dorsiflex and evert the foot.

Pitfalls

If the electrode is inserted too medially it will be in the extensor hallucis longus; if inserted too proximally it will be in the tibialis anterior or the extensor digitorum longus.

Comment

(a) This is the only peroneal muscle located in the anterior compartment of the leg. Therefore, it becomes involved in anterior compartment syndrome.

(b) This muscle may be considered as part of the extensor digitorum longus (its fifth tendon).

(c) It is involved in lesions of:

1. Deep peroneal nerve
2. Common peroneal nerve
3. Sciatic nerve
4. Sacral plexus
5. L5, S1.

POPLITEUS

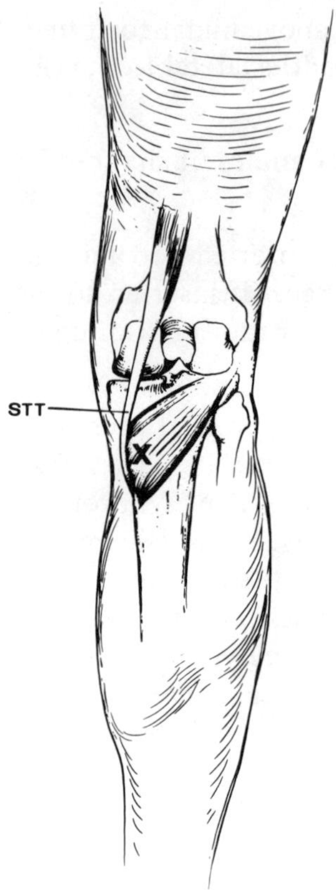

Innervation

Tibial Nerve, Sciatic Nerve, Anterior Division Sacral Plexus, *L5*, S1.

Origin

From the fibrous capsule of the knee joint on the side of the lateral condyle of the femur.

Insertion

Into the triangular area of the tibia above the soleal line.

Position

The patient prone with the knee fully extended.

Electrode Insertion (X)

The needle electrode is inserted laterally to the insertion of the semitendinosus tendon (STT).

Test Maneuver

Flex and internally rotate tibia.

Pitfalls

If the needle electrode is inserted either too distally or too superficially it will be in the gastrocnemius.

Comment

Involved in lesions of:

1. High tibial nerve
2. Sciatic nerve
3. Sacral plexus
4. L5, S1 roots.

SOLEUS

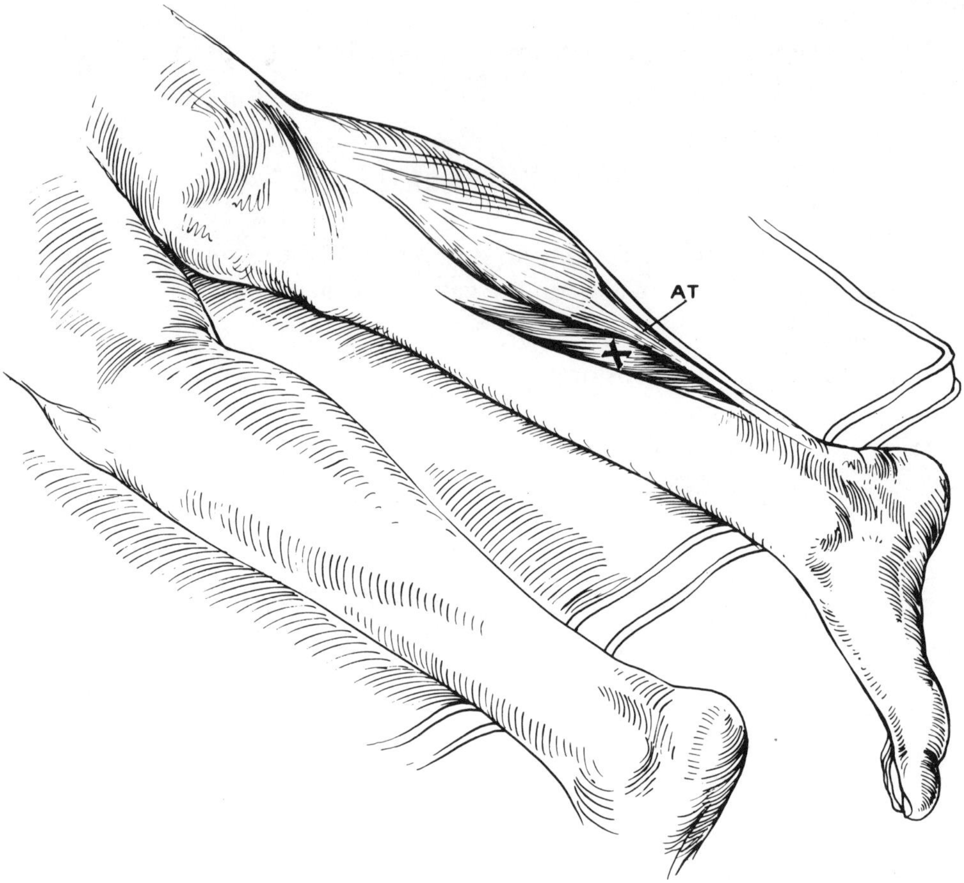

Innervation

Tibial Nerve, Sciatic Nerve, Anterior Division Sacral Plexus, L5, *S1, S2.*

Origin

From the head and the proximal portion of the body of the fibula and the middle one-third of the medial border of the tibia.

Insertion

Through the Achille's tendon, on calcaneous bone.

Electrode Insertion (X)

Insert the electrode distal to the belly of the gastrocnemius muscle, medial and anterior to the achilles tendon (AT).

Test Maneuver

Patient to plantar flex foot with knee flexed.

Pitfalls

If electrode is inserted too superficially it will be in the gastrocnemius.

Comments

(a) Muscle most commonly used to study the "H" reflex.

(b) This muscle is involved in lesions of:

1. Tibial nerve
2. Sciatic nerve
3. Sacral plexus
4. S1, S2 roots.

TIBIALIS ANTERIOR

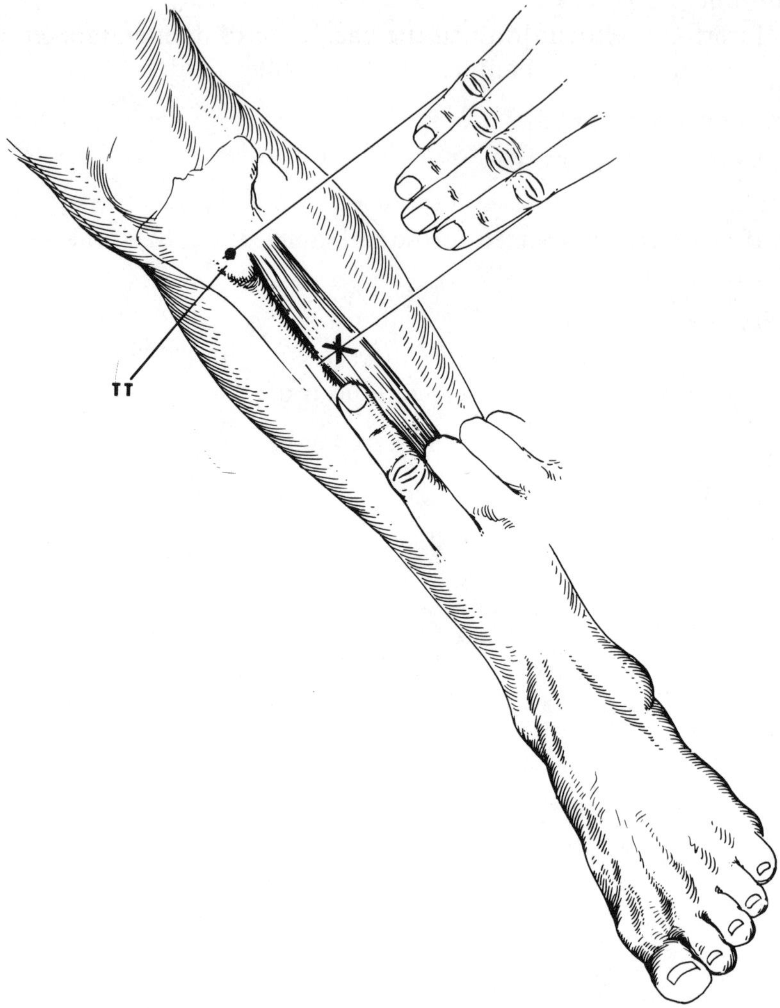

Innervation

Deep Peroneal Nerve, Common Peroneal Nerve, Sciatic Nerve, Posterior Division Sacral Plexus, L4, L5.

Origin

From the lateral condyle and the proximal two-thirds of the shaft of the tibia.

Insertion

On the first cuneiform and the base of the first metatarsal.

Position

The patient supine.

Electrode Insertion (X)

Four fingerbreadths below the tibial tuberosity (TT) and one fingerbreadth lateral to the tibial crest.

Test Maneuver

Patient to dorsiflex the foot.

Pitfalls

If the electrode is inserted too laterally and too deeply it will be in the extensor digitorum communis.

Comments

(a) First muscle innervated by the deep peroneal nerve.

(b) Involved in lesions of:

1. Deep peroneal nerve
2. Common peroneal nerve
3. Sciatic nerve
4. Sacral plexus
5. L4, L5 roots.

TIBIALIS POSTERIOR

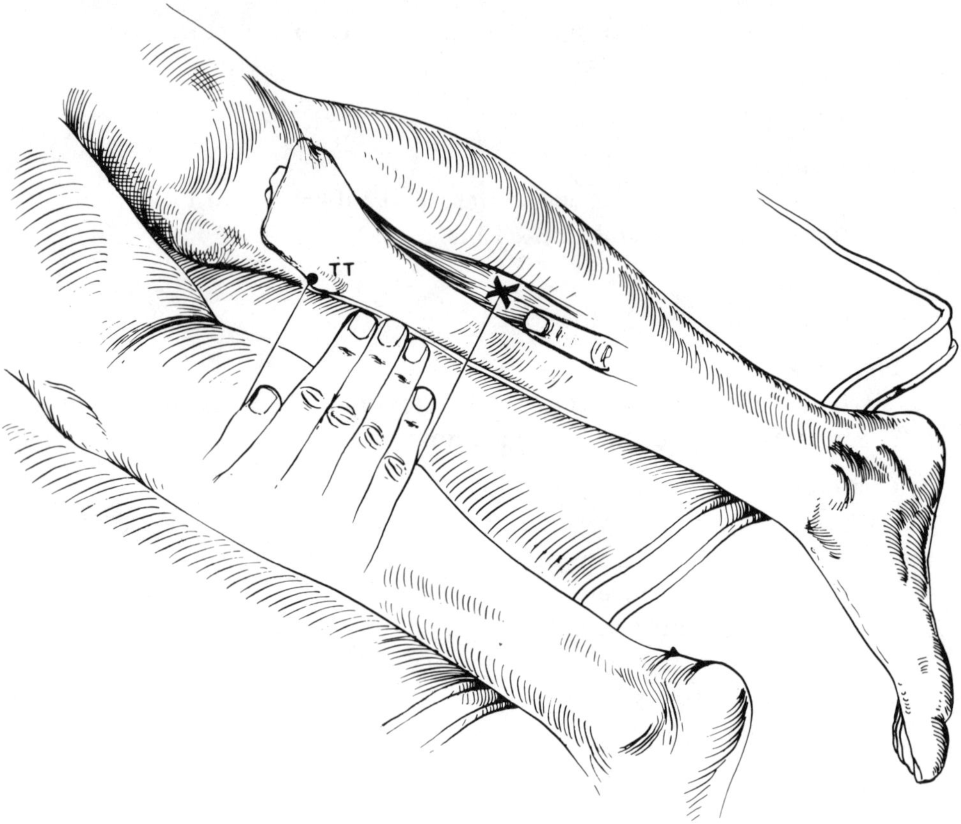

Innervation

Tibial Nerve, Sciatic Nerve, Anterior Division Sacral Plexus, L5, S1.

Origin

From the interosseus membrane, the posterior surface of the body of the tibia and the upper two-thirds of the medial surface of the fibula.

Insertion

On the tuberosity of the navicula bone and through fibrous expansions into the sustentaculumtali and the calcaneous.

Position

The patient prone with feet over edge of plinth, thigh internally rotated.

Electrode Insertion (X)

One handbreadth distal to the tibial tuberosity (TT) and one fingerbreadth off the medial edge of the tibia. The electrode is directly obliquely through the soleus and flexor digitorum longus, just posterior to the tibia.

Test Maneuver

Patient to invert foot in plantar flexion.

Pitfalls

If the electrode is inserted too superficially it will be in the soleus or flexor digitorum longus; if inserted too deeply it will be in the tibialis anterior.

Comment

Involved in lesions of:

1. Tibial nerve
2. Sciatic nerve
3. Sacral plexus
4. L5, S1 root.

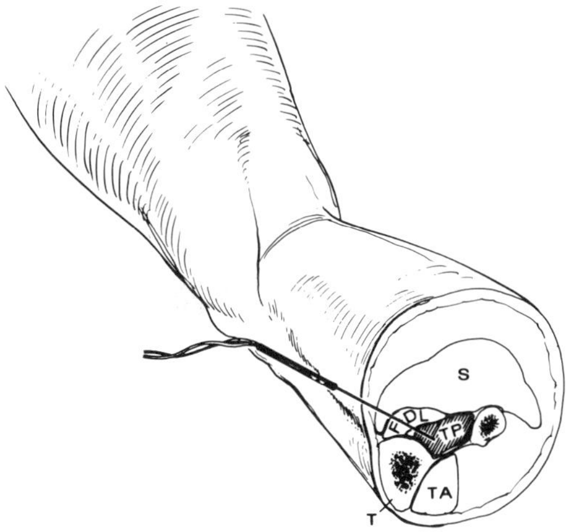

TP–Tibialis Posterior
FDL–Flexor Digitorum Longus
S–Soleus
T–Tibia
TA–Tibialis Anterior

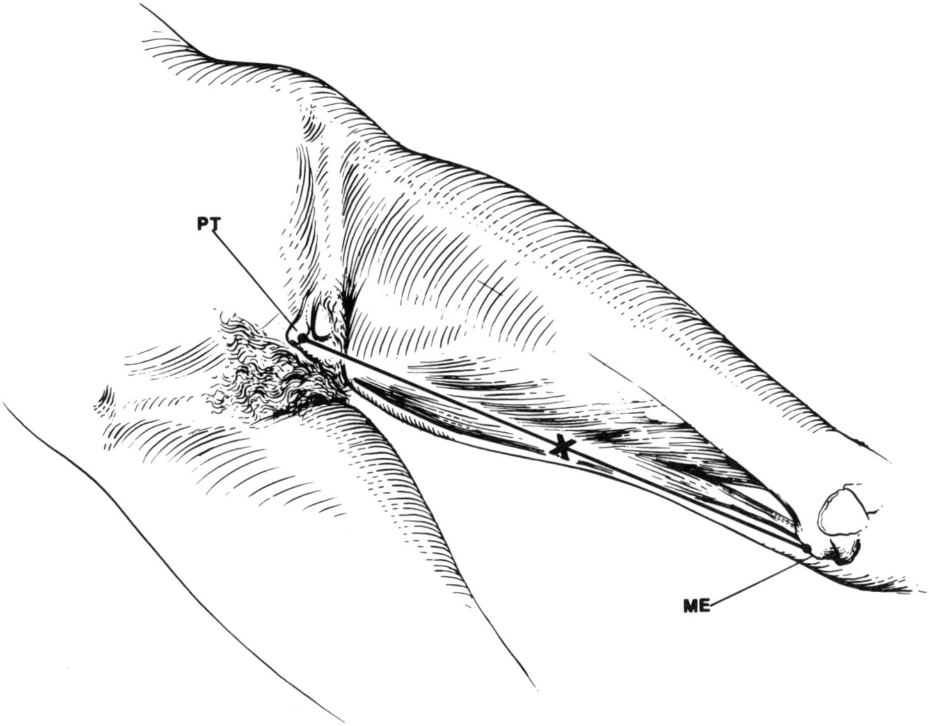

SECTION VIII
THIGH

ADDUCTOR BREVIS

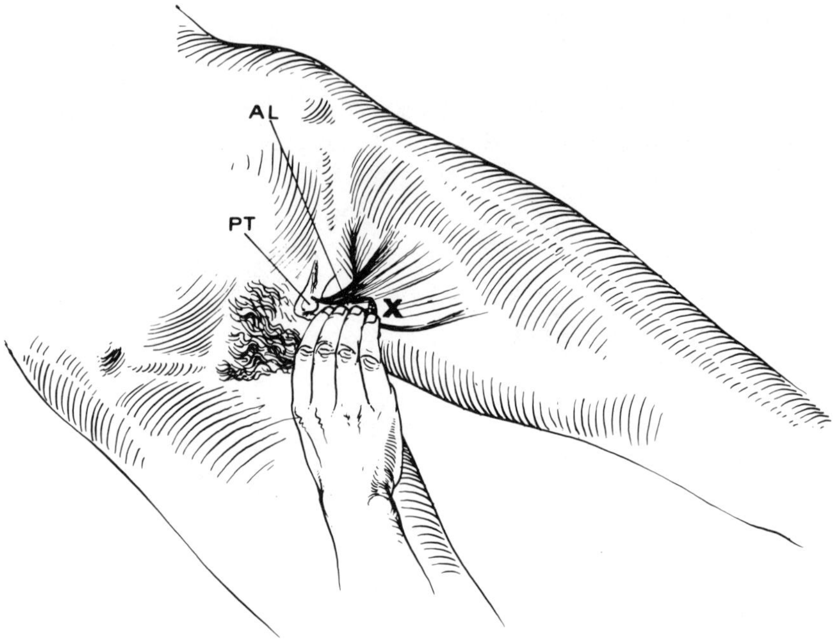

Innervation

Obturator Nerve, Anterior Division Lumbar Plexus, L2, L3, L4.

Origin

From the inferior ramus of the pubis.

Insertion

To a line below the lesser trochanter of the femur.

Position

The patient supine with both lower extremities abducted fifteen degrees.

Electrode Insertion (X)

Palpate the tendon of adductor longus (AL) arising from pubic tubercle (PT) and insert electrode four fingerbreadths distal to tubercle through the adductor longus muscle to about two inches.

Text Maneuver

Patient to adduct the limb.

Pitfalls

If the electrode is inserted too superficially it will be in the adductor longus (laterally) or in the gracilis (medially); if inserted too medially it will be in the adductor magnus.

Comment

Involved in–

1. Anterior branch of the obturator nerve lesions
2. Obturator nerve lesions
3. Anterior division of the lumbar plexus lesions
4. L2, L3, L4 roots lesions
5. Landmark for obturator nerve block. Anterior branch lies on ventral surface of muscle, posterior branch lies on the posterial surface.

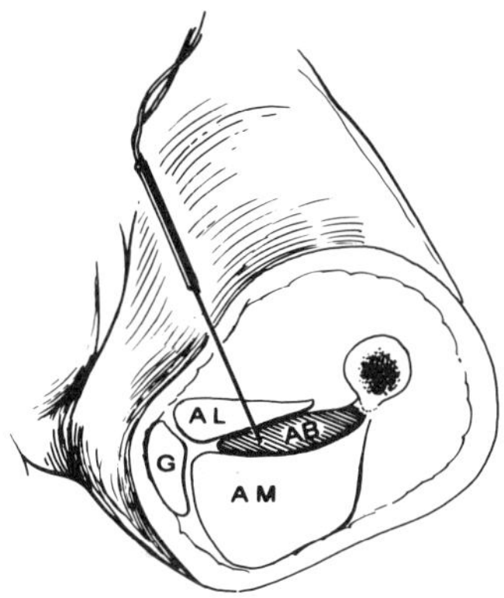

AB–Adductor Brevis
AL–Adductor Longus
AM–Adductor Magnus
G–Gracilis

ADDUCTOR LONGUS

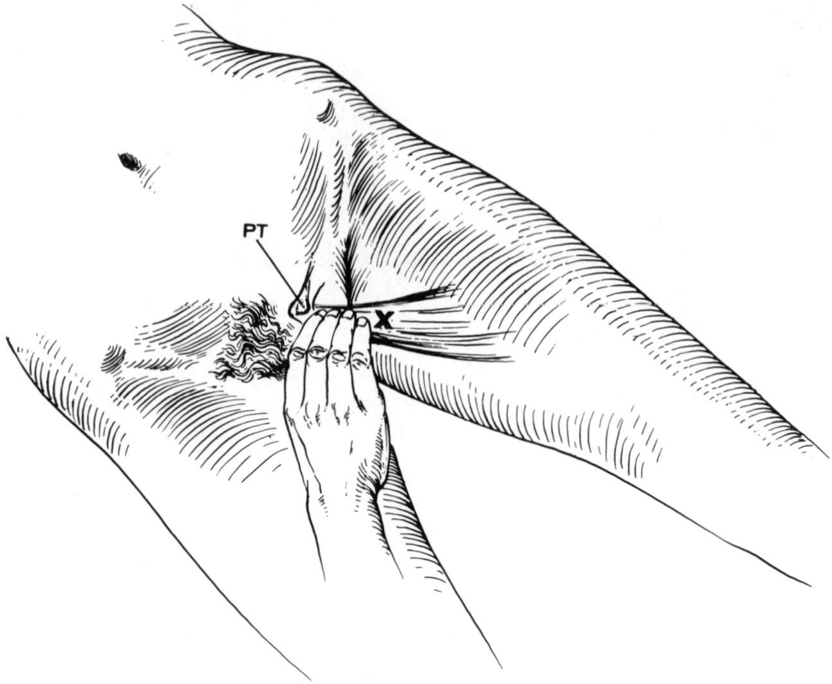

Innervation

Obturator Nerve, Anterior Division Lumbar Plexus, *L2, L3, L4*

Origin

From the pubic tubercle.

Insertion

Into the linea aspera, between the adductor magnus and the vastus medialis.

Position

The patient supine with both lower extremities abducted fifteen degrees.

Electrode Insertion (X)

Palpate the tendon arising from the pubic tubercle (PT) and insert the electrode four fingerbreadths distal to the pubic tubercle into the muscle belly.

Test Maneuver

Patient to adduct limb.

Pitfalls

If the electrode is inserted too medially it will be in the gracilis; if inserted too laterally it will be in the sartorius.

Comment

Involved in–

1. Anterior branch of the obturator nerve lesions
2. Obturator nerve lesions
3. Anterior division of the lumbar plexus lesions
4. L2, L3, L4 roots lesions.

ADDUCTOR MAGNUS

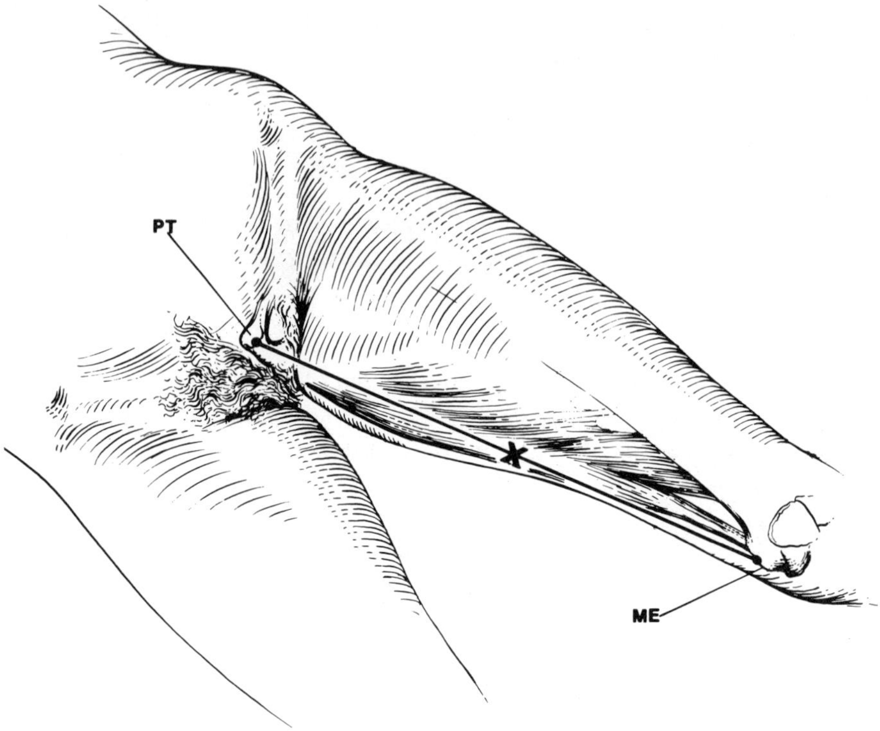

Innervation

Obturator Nerve, Anterior Division Lumbar Plexus, L2, L3, L4, L5.

Also Fibers from sciatic nerve L4, L5.

Origin

From the inferior ramus of the pubis and ischium and the tuberosity of the ischium.

Insertion

Into the linea aspera and the adductor tubercle of the femur.

Position

The patient supine with both lower extremities abducted fifteen degrees and externally rotated.

Electrode Insertion (X)

Midway between the medial femoral epicondyle (ME) and the pubic tubercle (PT).

Test Maneuver

Patient to adduct the thigh.

Pitfalls

If the electrode is too superficial it will be in the gracilis; if inserted too laterally it will be in the sartorius; and if inserted too proximally it will be in the adductor longus.

Comments

(a) Partially involved in–

1. Posterior branch of the obturator nerve[4]
2. Obturator nerve lesions
3. Anterior division of the lumbar plexus
4. L2, L3, L4 roots lesions.

(b) Also involved in–

1. High lesions of the sciatic nerve
2. Anterior divisions of the upper portion of the sacral plexus
3. L4, L5 roots lesions.

[4] Morphologically the superficial portion of the muscle is considered to be part of the hamstrings muscle group. This accounts for its innervation through the sciatic nerve.

BICEPS FEMORIS: Long Head

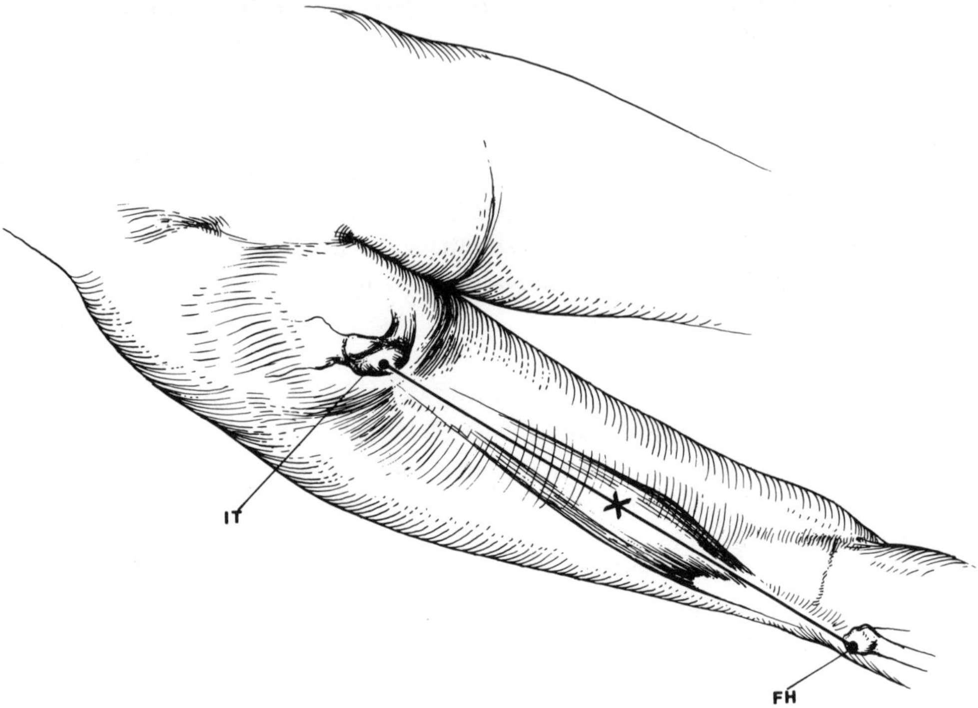

Innervation

Sciatic Nerve (Tibial Portion), Anterior Division, Sacral Plexus, L5, *S1*.

Origin

From the ischial tuberosity.

Insertion

Into the head of the fibula.

Position

The patient prone.

Electrode Insertion (X)

Insert the electrode at the midpoint of a line between the fibula head (FH) and the ischial tuberosity (IT).

Test Maneuver

Patient to flex the knee.

Pitfalls

None.

Comment

Involved in lesions of–

1. Sciatic nerve
2. Anterior division of the sacral plexus
3. L5, S1 roots lesion.

BICEPS FEMORIS: Short Head

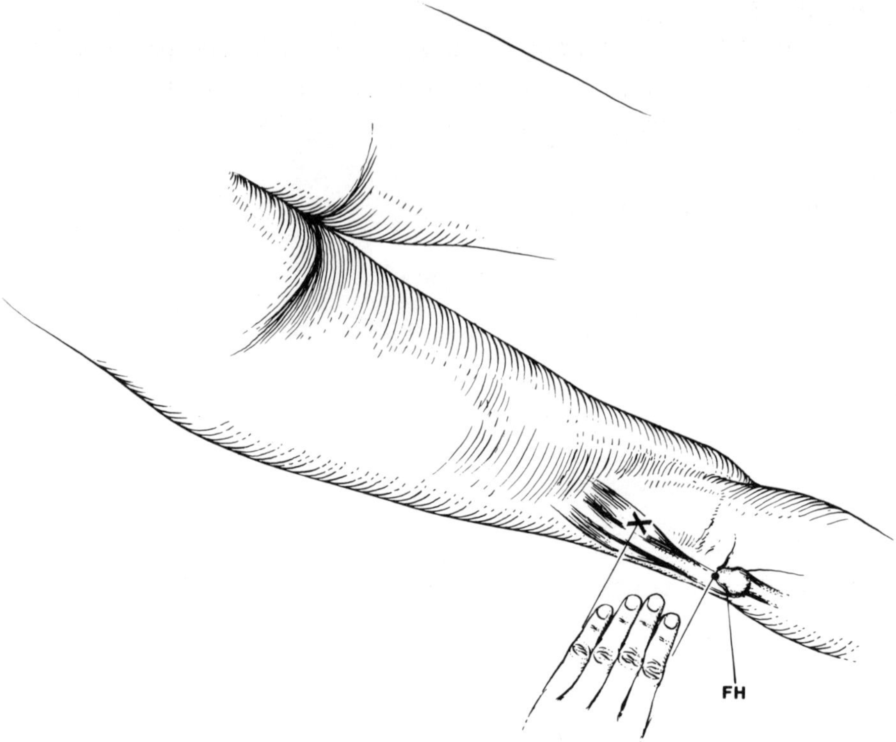

Innervation

Sciatic Nerve, (Peroneal Division) Posterior Division, Sacral Plexus, L5, *S1*, *S2*.

Origin

From the lateral lip of the linea aspera and the upper part of the lateral supracondylar line.

Insertion

Into the head of the fibula and the lateral condyle of the tibia.

Position

The patient prone with the knee flexed to ninety degrees.

Electrode Insertion (X)

Palpate the tendon of the long head of the biceps; grasp it with the fingertips; insert the electrode just medial to it, four fingerbreadths proximal to the fibular head (FH).

Test Maneuver

Patient to flex the knee.

Pitfalls

If the electrode is inserted too medially it will be in the semimembranosus; if inserted too laterally it will be in the long head of the biceps femoris.

Comment:

(a) Involved in lesions of –
   1. Sciatic Nerve
   2. Anterior division of the sacral plexus
   3. L5, S1 roots lesion.

(b) Only muscle above the knee innervated by the peroneal division of the sciatic nerve.

GRACILIS

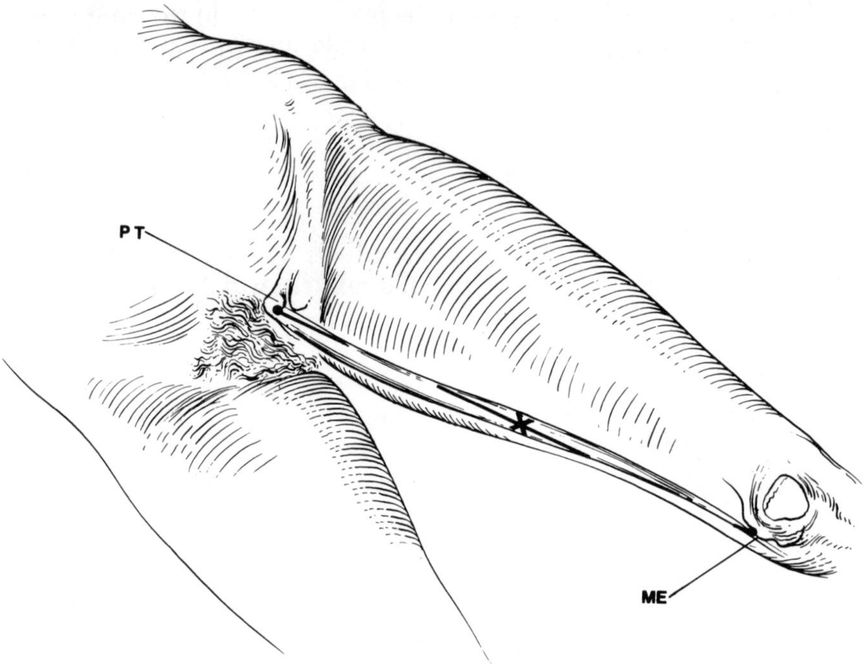

Innervation

Obturator Nerve, Anterior Division Lumbar Plexus, *L2, L3,* L4.

Origin

From the lower half of the symphysis pubis and the pubic arch.

Insertion

Into the medial surface of the tibia below the condyle.

Electrode Insertion (X)

Insert the electrode to a maximum depth of one-half inch at a point midway between the pubic tubercle (PT) and the medial femoral epicondyle (ME).

Test Maneuver

Patient to adduct the thigh and flex the knee.

Pitfall

If the electrode is inserted too deeply it will be in the adductor magnus; if inserted too laterally it will be in the adductor longus.

Comment

Involved in lesions of–

1. Anterior branch of the obturator nerve
2. Obturator for nerve
3. Anterior division of the lumbar plexus
4. L2, L3, L4 roots lesions.

ILIOPSOAS

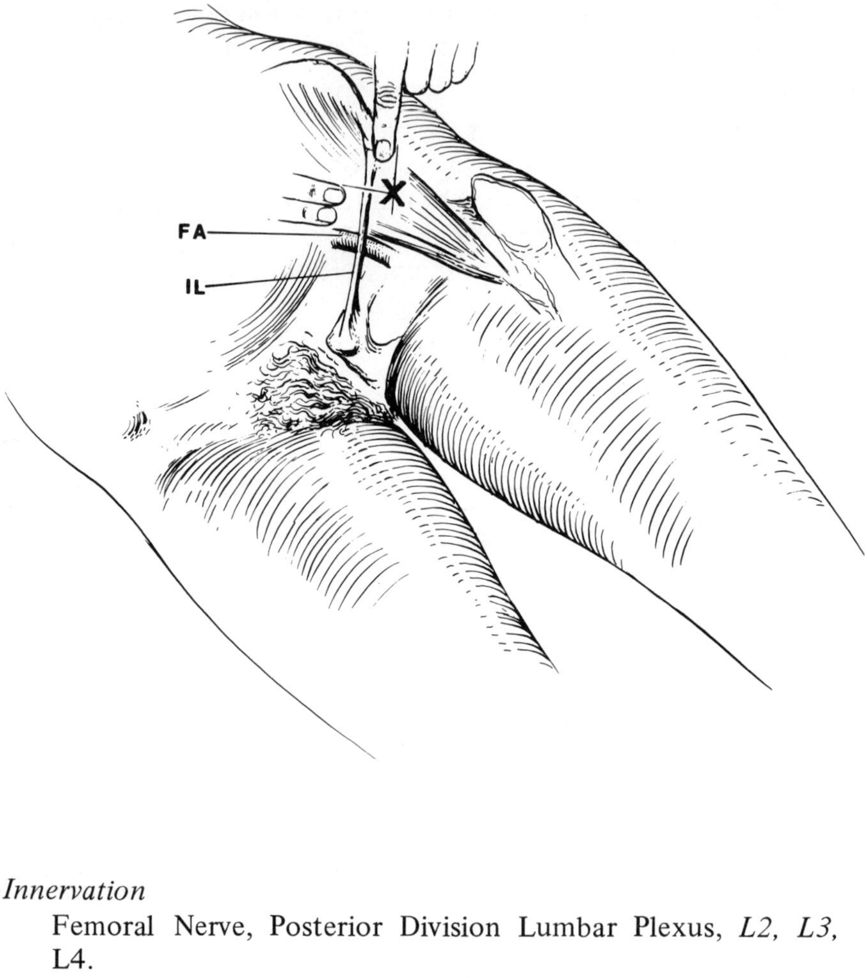

Innervation

Femoral Nerve, Posterior Division Lumbar Plexus, *L2, L3,* L4.

Origin

From the bodies and the transverse processes of the lumbar vertebrae and the iliac fossa.

Insertion

Into the lesser trochanter of the femur.

Position

The patient supine.

Electrode Insertion (X)

Two fingerbreadths lateral to the femoral artery (FA) and one fingerbreadth below the inguinal ligament (IL).

Test Maneuver

Patient to flex the thigh with the knee flexed beyond ninety degrees.

Pitfalls

If the electrode is inserted too medially it will contact the neurovascular bundle; if inserted too laterally it will be in the sartorius.

Comment

(a) Involved in lesions of–
   1. High femoral nerve
   2. Posterior Division of the lumbar plexus
   3. L2, L3, L4 roots.

(b) Forms the external portion of the floor of the Scarpa's triangle.

PECTINEUS

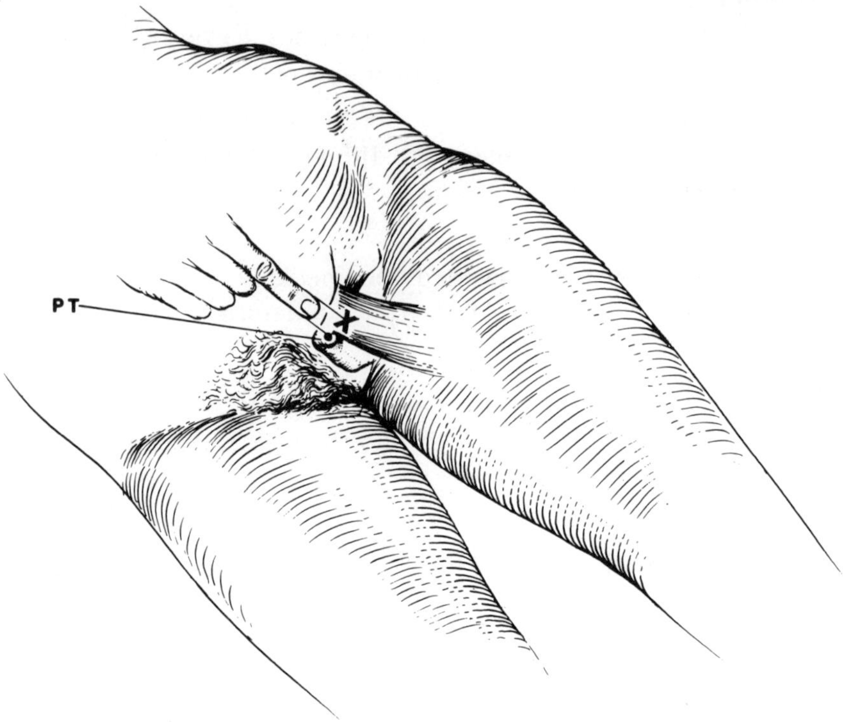

Innervation

Femoral Nerve, Posterior Division Lumbar Plexus, *L2, L3,* L4.

Origin

From the superior ramus of the pubis.

Insertion

Into the upper portion of the pectineal line below the lesser trochanter.

Position

The patient supine.

Electrode Insertion (X)

One fingerbreadth lateral to the pubic tubercle (PT).

Maneuver

Patient to adduct the thigh.

Pitfalls

If the electrode is inserted too medially it will be in the adductor longus; if inserted too laterally it will contact the neurovascular bundle.

Comment

(a) Forms the internal portion of the floor of the Scarpa's triangle.

(b) Involved in lesions of–

1. Femoral nerve (entrapment) at the inguinal ligament level
2. Femoral nerve proximal to the inguinal ligaments
3. Posterior division of the lumbar plexus
4. L2, L3, L4 roots.

RECTUS FEMORIS

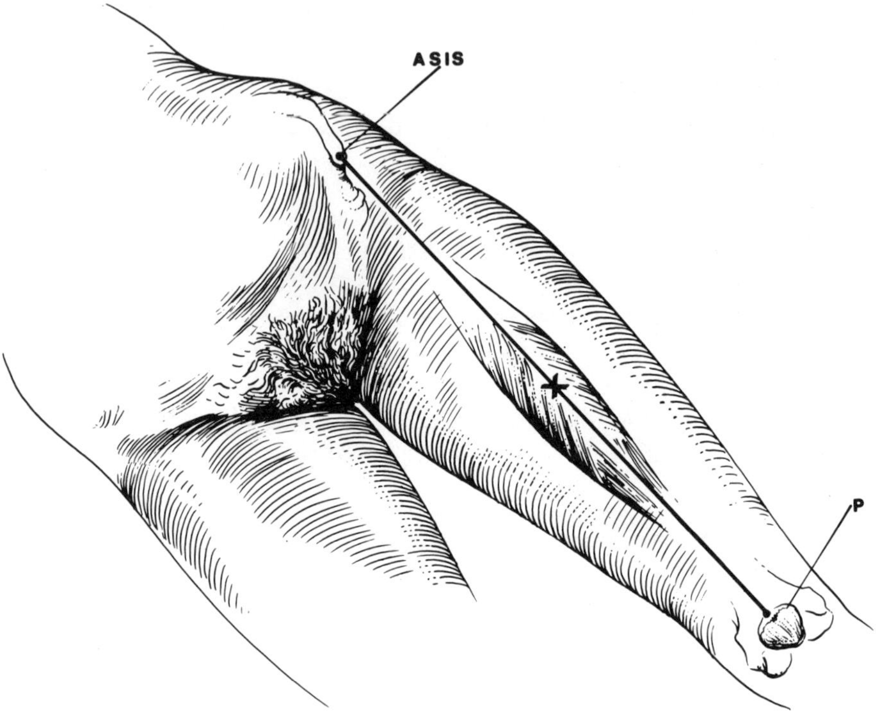

Innervation

Femoral Nerve, Posterior Division Lumbar Plexus, L2, *L3, L4.*

Origin

From the anterior inferior iliac spine and the brim of the acetabulum.

Insertion

Through the patella tendon, on the tibial tubercle.

Position

The patient supine.

Electrode Insertion (X)

On the anterior aspect of the thigh, midway between the superior border of the patella (P) and the anterior superior iliac spine (ASIS).

Test Maneuver

Patient to flex the hip with the knee extended.

Pitfalls

If the electrode is inserted too medially it will be in the vastus intermedius; if inserted too laterally it will be in the vastus lateralis; if inserted too distally and medially it will be in the vastus medialis.

Comment

Involved in lesions of–

1. Femoral nerve (entrapment) at the inguinal ligament level
2. Femoral nerve proximal to the inguinal ligament
3. Posterior division of the lumbar plexus
4. L2, L3, L4.

SARTORIUS

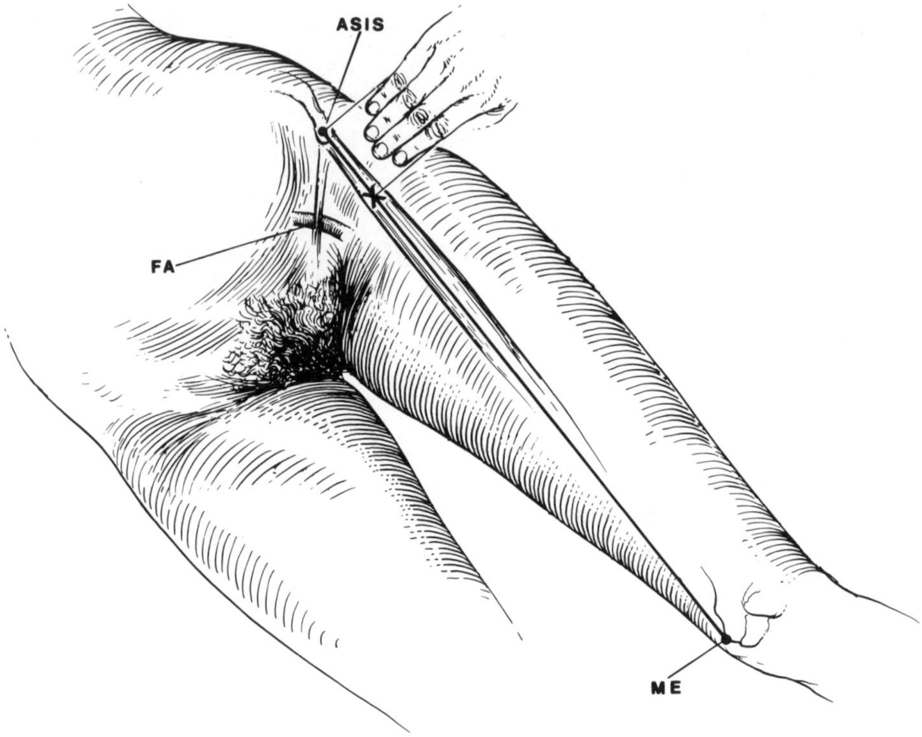

Innervation

Femoral Nerve, Posterior Division Lumbar Plexus, *L2, L3, L4.*

Origin

From the anterior superior iliac spine.

Insertion

Below the medial tibial condyle, on the medial border of the body of the tibia.

Position

The patient supine.

Electrode Insertion (X)

Four fingerbreadths distal to the anterior superior iliac spine (ASIS) along the line to the medial epicondyle (ME) of the tibia. Insert the electrode just lateral to the femoral artery (FA) to a depth of about one-half inch.

Test Maneuver

Patient to flex, abduct and externally rotate thigh.

Pitfalls

If the needle electrode is inserted too deeply or too distally it will be in the rectus femoris; if inserted too medially it will be in the iliacus; if inserted too laterally it will be in the tensor fascia lata.

Comment

Involved in lesions of–

1. Femoral nerve (entrapment) at the inguinal ligament level
2. Femoral nerve proximal to the inguinal ligament
3. Posterior division of the lumbar plexus
4. L2, L3, L4.

SEMIMEMBRANOSUS

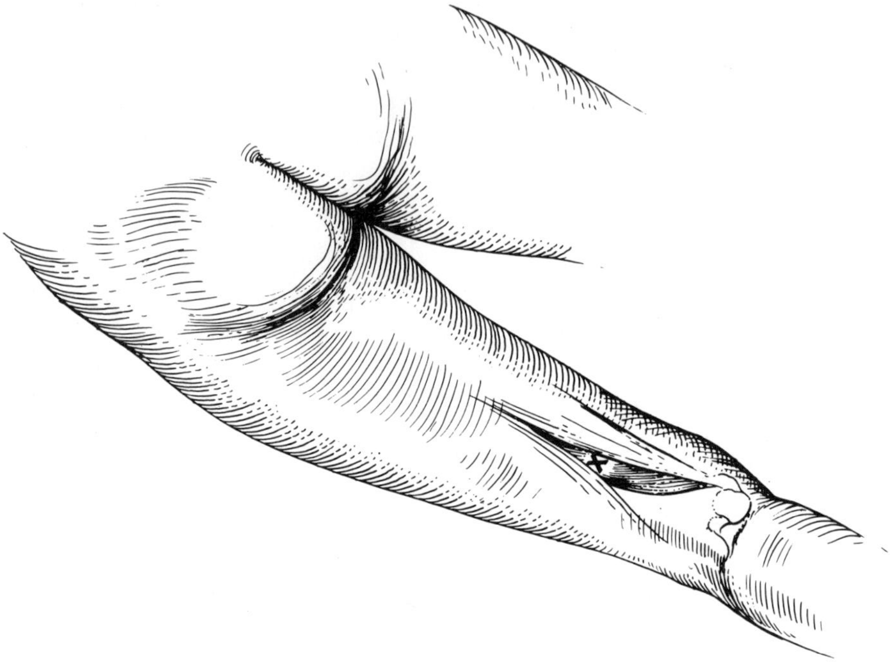

Innervation

Sciatic Nerve (Tibial Portion), Anterior Division Sacral Plexus, L5, S1, S2.

Origin

From the ischial tuberosity.

Insertion

On the medial condyle of the tibia and through a fibrous expansion into the lateral femoral condyle.

Position

The patient prone.

Electrode Insertion (X)

Insert the electrode laterally to the semitendinosus tendon in the apex of the "V" between the semitendinosus tendon and the biceps femoris.

Test Maneuver

Patient to flex the knee and internally rotate the tibia.

Pitfalls

If the electrode is inserted too medially it will be in the semitendinosus; if inserted too laterally it will be the short head of the biceps or into the sciatic nerve; if inserted too deeply it will be in the adductor magnus.

Comment

Involved in lesions of–

1. Sciatic nerve
2. Anterior division of the sacral plexus
3. L5, S1, S2.

SEMITENDINOSUS

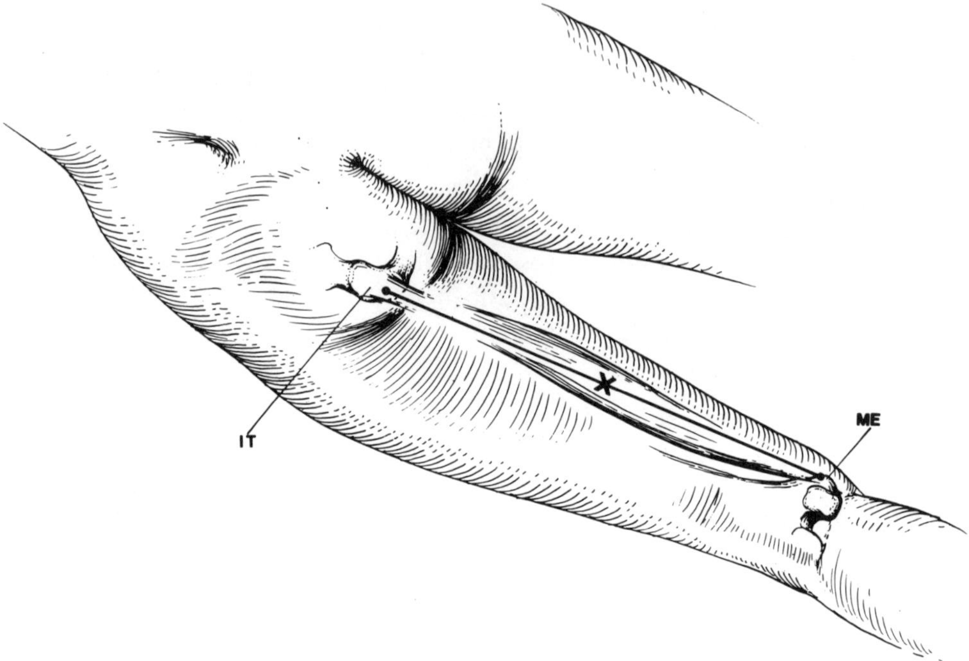

Innervation

Sciatic Nerve (Tibial Portion), Anterior Division Sacral Plexus, L5, S1, S2.

Origin

From the ischial tuberosity.

Insertion

On the medial condyle of the tibia.

Position

The patient prone.

Electrode Insertion (X)

Midway on a line between the medial epicondyle (ME) of the femur and the ischial tuberosity (IT).

Test Maneuver

Patient to flex the knee and internally rotate the tibia;

Pitfalls

If the electrode is inserted too laterally it will be in the long head of the biceps; if inserted too medially or too deeply it will be in the semimembranosus.

Comment

Involved in lesions of–

1. Sciatic nerve
2. Anterior division sacral plexus
3. L5, S1, S2 roots.

TENSOR FASCIE LATAE

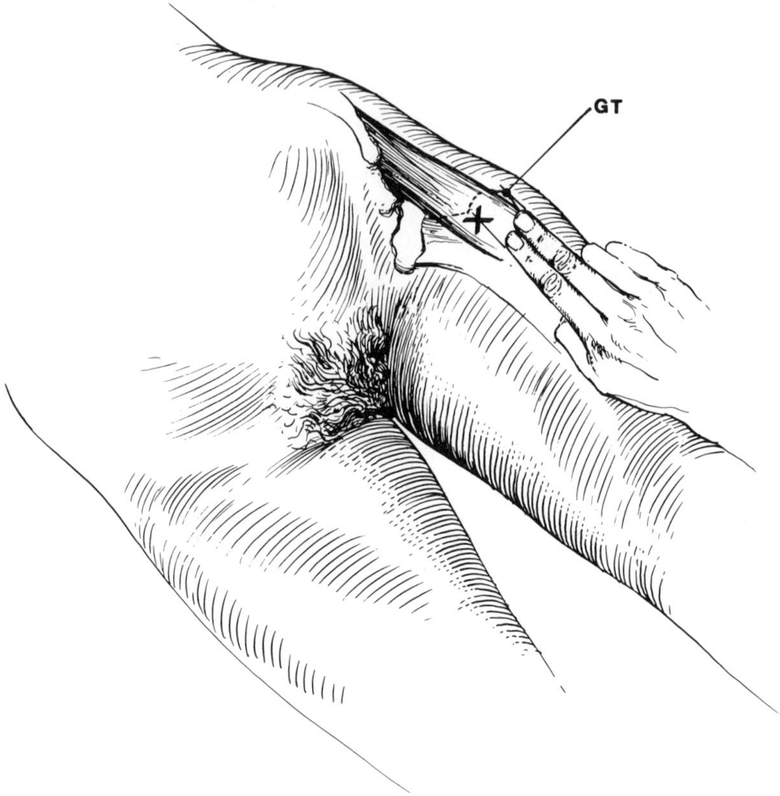

Innervation

Superior Gluteal Nerve, Sacral Plexus, L4, *L5, S1.*

Origin

From outer lip of the anterior portion of iliac crest of the ilium.

Insertion

On the iliotibial tract, just below the greater trochanter.

Position

The patient supine.

Electrode Insertion (X)

Two fingerbreadths anterior to the greater trochanter (GT).

Test Maneuver

Patient to abduct thigh with hip flexed.

Pitfalls

If the electrode is inserted too anteriorly it will be in the sartorius or rectus femoris; if inserted too deeply it will be in the vastus lateralis; if inserted too posteriorly it will be in the gluteus medius.

Comment

Involved in lesions of–

1. Superior gluteal nerve
2. Posterior division of the sacral plexus
3. L4, L5, S1 roots.

VASTUS INTERMEDIUS

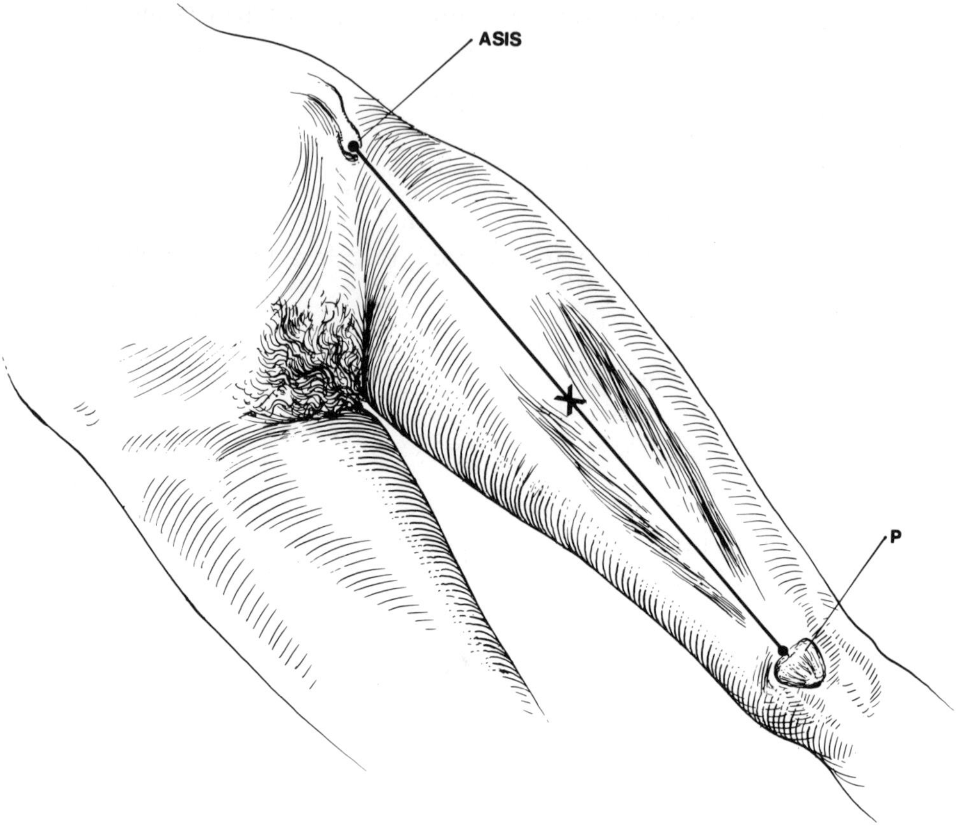

Innervation

Femoral Nerve, Posterior Division Lumbar Plexus, *L2, L3, L4.*

Origin

From the upper three-fourths of the shaft of the femur and from the anterior surface as high as the introchanteric line.

Insertion

Through the quadriceps tendon onto the tibial tubercle.

Position

The patient supine.

Electrode Insertion

Midway between the superior border of the patella (P) and the anterior superior iliac spine (ASIS). The electrode is inserted to the bone and withdrawn slightly.

Test Maneuver

Patient to lift heel from plinth with knee extended.

Pitfalls

If the electrode is inserted too superficially it will be in the rectus femoris; if inserted too laterally it will be in the vastus lateralis; if inserted too medially it will be in the vastus medialis or sartorius.

Comment

Involved in lesions of–

1. Femoral nerve (entrapment) at the inguinal ligament level
2. Femoral nerve proximal to the inguinal ligament
3. Posterior division of the lumbar plexus
4. L2, L3, L4 roots.

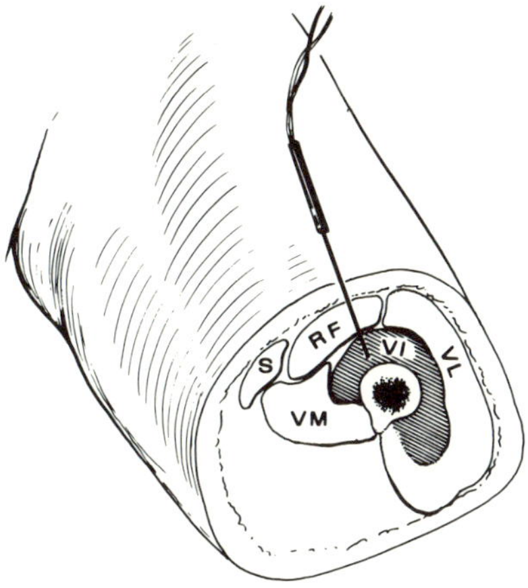

VI–Vastus Intermedius
VM–Vastus Medialis
VL–Vastus Lateralis
S–Sartorius
RF–Rectus Femoris

VASTUS LATERALIS

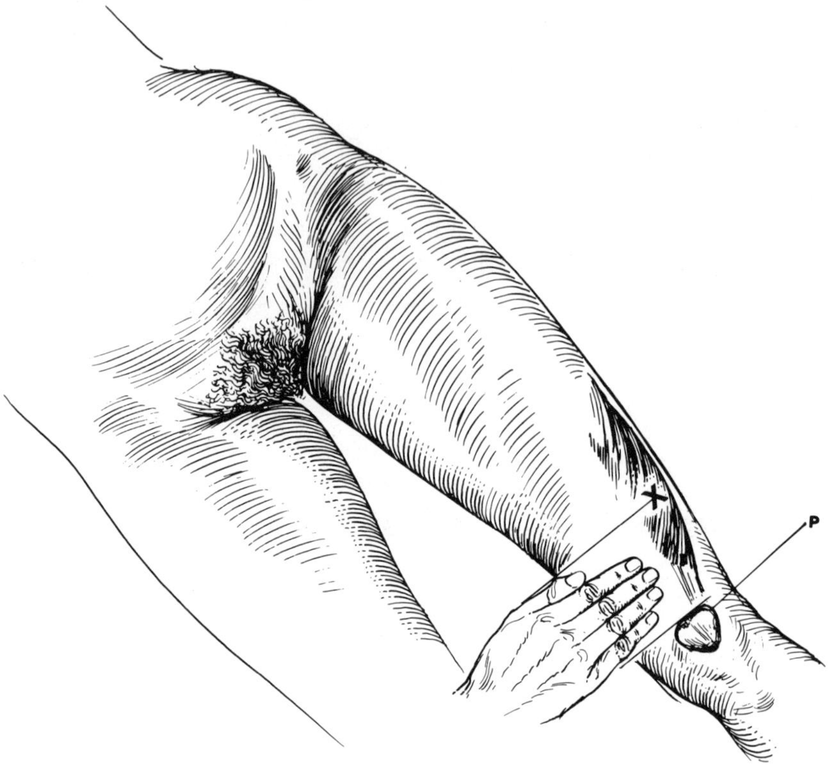

Innervation

Femoral Nerve, Posterior Division Lumbar Plexus, L2, *L3, L4.*

Origin

From the intertrochanteric line, the linea aspera and the medial sypracondylar line.

Insertion

Through the quadriceps tendon onto the tibial tubercle.

Position

The patient supine.

Electrode Insertion (X)

Over the lateral aspect of the thigh, one handbreadth above the patella (P).

Test Maneuver

Patient to lift heel from plinth with knee extended.

Pitfalls

If the electrode is inserted too posteriorly it will be in the biceps femoris; if inserted too medially it will be in the rectus femoris.

Comment

Involved in lesions of–

1. Femoral nerve (entrapment) at the inguinal ligament level
2. Femoral nerve proximal to the inguinal ligament
3. Posterior division of the lumbar plexus
4. L2, L3, L4 roots.

VASTUS MEDIALIS

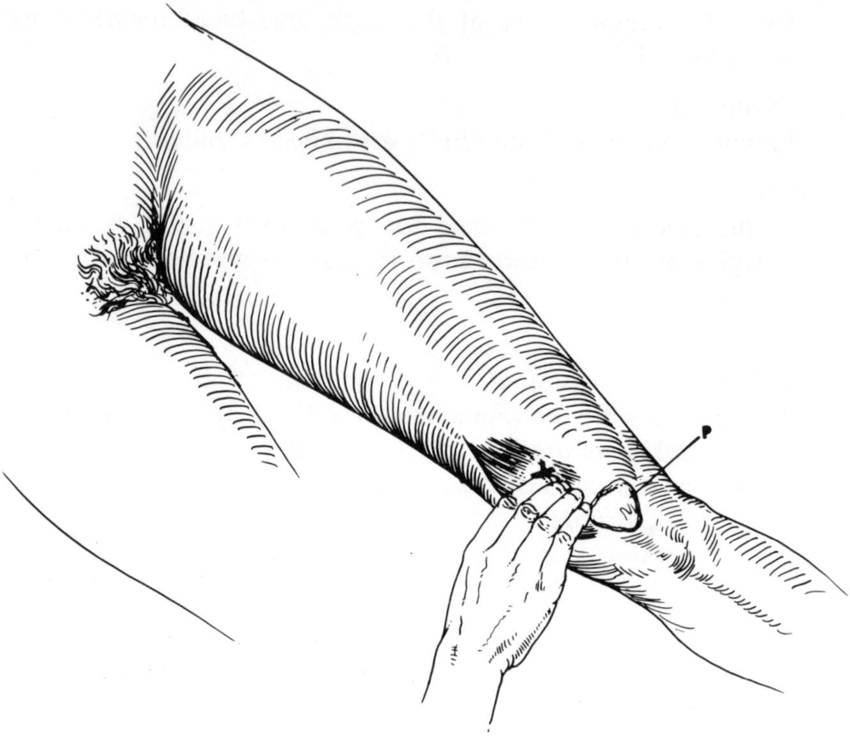

Innervation

Femoral Nerve, Posterior Division Lumbar Plexus, L2, *L3, L4.*

Origin

From the medial lip of the linea aspera and upper part of the supracondylar line.

Insertion

Through the quadriceps tendon onto the tibial tubercle.

Position

The patient supine.

Electrode Insertion (X)

Four fingerbreadths proximal to the superiomedial angle of the patella (P).

Test Maneuver

Patient to lift heel from plinth with knee extended.

Pitfalls

If the electrode is inserted too posteriorly it will be in the sartorius or gracilis; if inserted too anteriorly it will be in the rectus femoris.

Comment

(a) Used as recording muscle for femoral nerve motor conduction study

(b) Involved in lesions of–

1. Femoral nerve (entrapment) at the inguinal ligaments
2. Intrapelvic femoral nerve
3. Posterior division of lumbar plexus
4. L2, L3, L4 nerve roots.

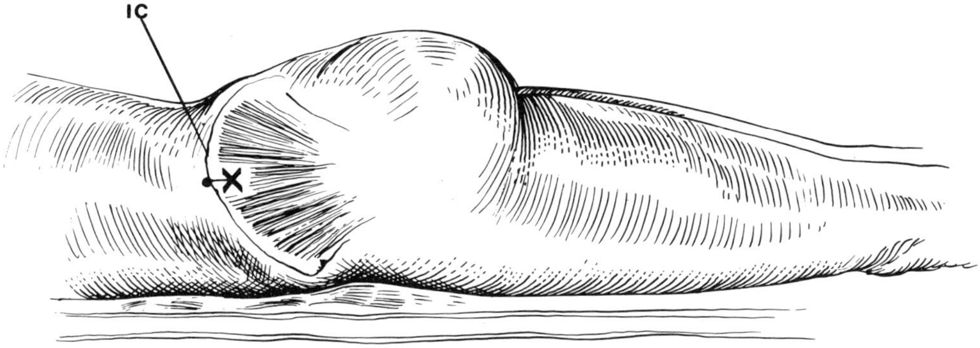

SECTION IX

PELVIS AND HIP JOINT

GLUTEUS MAXIMUS

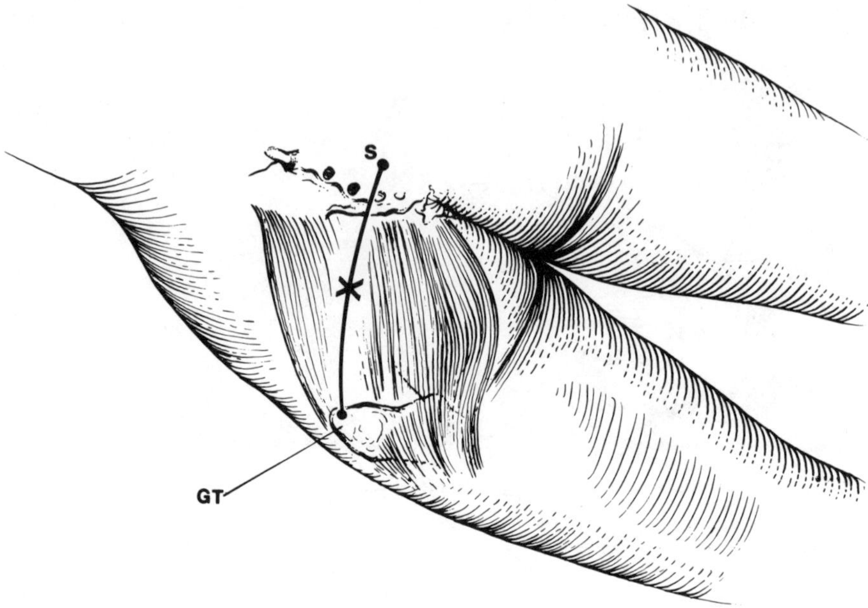

Innervation

Inferior Gluteal Nerve, Sacral Plexus, L5, *S1, S2.*

Origin

From the gluteal line, the posterior surface of the sacrum and coccyx, and the sacrotuberous ligament.

Insertion

Into the iliotibial tract, the gluteal ridge, and the linea aspera of the femur.

Position

The patient prone.

Electrode Insertion (X)

The electrode is inserted to a depth of one to three inches midway between the greater trochanter (GT) and the sacrum (S).

Test Maneuver

Patient to extend the hip with the knee flexed.

Pitfalls

None.

Comment

(a) Involved in lesions of–
  1. Inferior gluteal nerve
  2. Posterior division of sacral plexus
  3. L5, S1, S2 nerve roots.

(b) A frequent site of intramuscular injection therefore, pathological findings may be misleading.

GLUTEUS MEDIUS

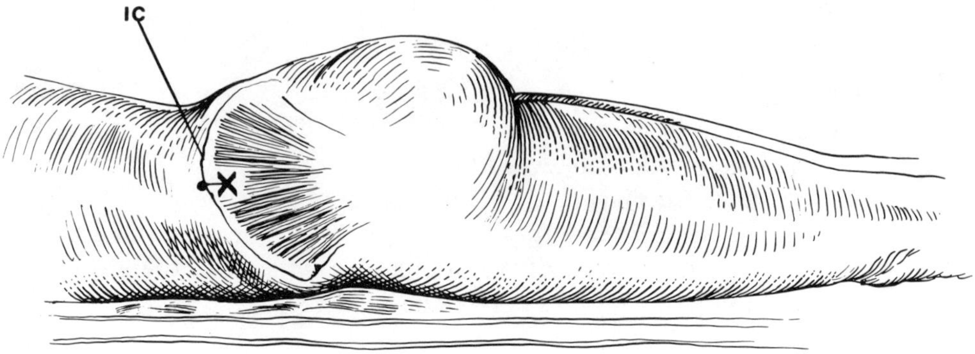

Innervation

Superior Gluteal Nerve, L4, L5, S1.

Origin

From the anterior gluteal line and the crest of the ilium.

Insertion

Into the greater trochanter.

Position

The patient prone.

Electrode Insertion (X)

One inch distal to the midpoint of the iliac crest (IC).

Test Maneuver

Patient to abduct the thigh.

Pitfalls

If the electrode is inserted too posteriorly it will be in the gluteus maximus; if inserted too anteriorly it will be in the tensa fascia lata; if inserted too distally it will be in either the gluteus minimus or maximus.

GLUTEUS MINIMUS

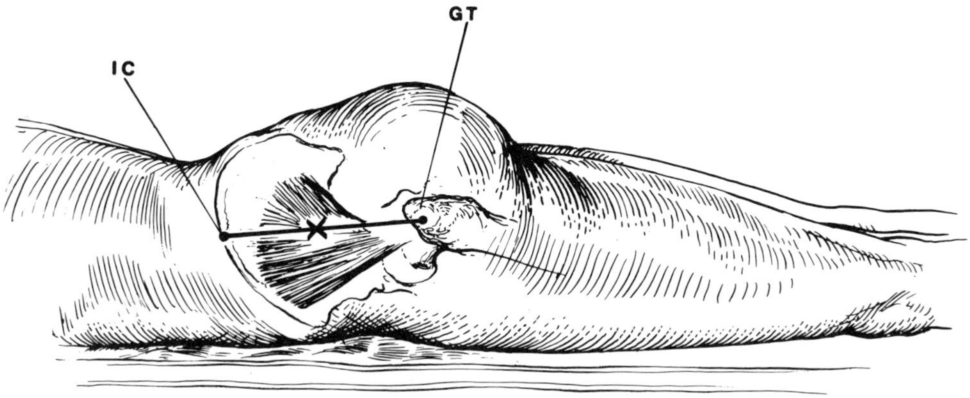

Innervation

Superior Gluteal Nerve, Sacral Plexus, L4, *L5*, S1.

Origin

From the lateral surface of the ilium, between the anterior and inferior gluteal line.

Insertion

Into the anterior surface of the greater trochanter.

Position

The patient prone.

Electrode Insertion (X)

Insert electrode midway between midpoint of the iliac crest (IC) and greater trochanter (GT) deep to bone and withdraw slightly.

Test Maneuver

Patient to abduct the thigh.

Pitfalls

If the electrode is inserted too superficially it will be in the gluteus medius.

OBTURATOR INTERNUS AND GEMELLI[5]

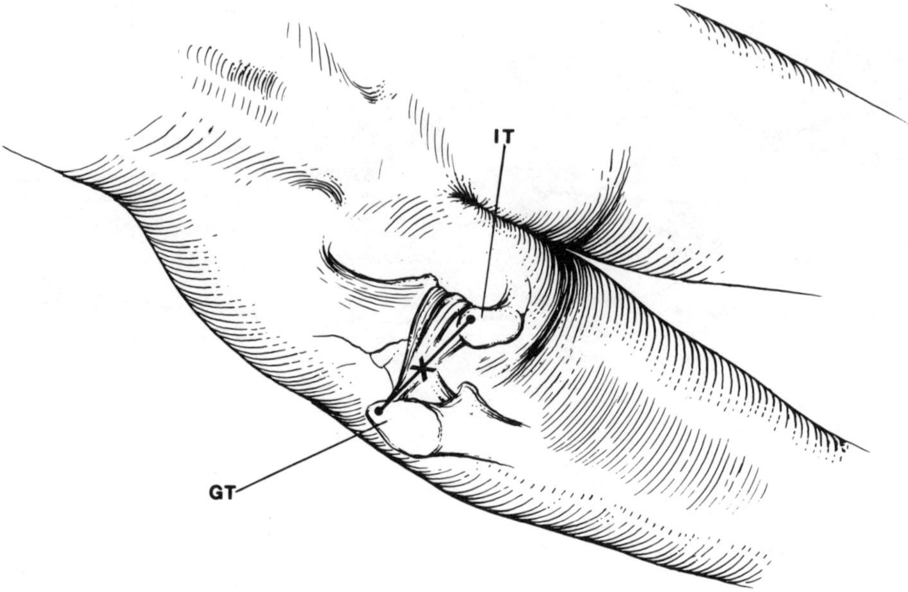

Innervation

Obturator Internus and Superior Gemellus: Obturator Internus Nerve, Sacral Plexus, L5, *S1, S2.*

Inferior Gemellus: Quadratus Femoris Nerve, Sacral Plexus L4, *L5,* S1, S2.

Both Nerves: Arise From the Proximal Segment of the Tibial Portion of the Sciatic Nerve

Origin

From the ischial spine, the ischial tuberosity, the obturator membrane and the adjacent bone.

Insertion

Into the medial side of the greater trochanter.

Position

The patient prone.

[5] The gemelli are considered the extra pelvic portion of the obturator internus.

Electrode Insertion

Insert electrode to bone and withdraw slightly at a point midway between the ischial tuberosity (IT) and the posterosuperior aspect of greater trochanter (GT).

Test Maneuver

Patient to externally rotate the thigh with the leg extended.

Pitfalls

If the electrode is inserted too medially it will contact the sciatic nerve; if inserted too superficially it will be in the gluteus maximus; if inserted too distally it will be in the quadratus femoris; if inserted too proximally it will be in the piriformis.

Comment

Involved in lesions of the lumbosacral plexus close to the spine or in L5, S1, S2 radiculopathies.

PIRIFORMIS

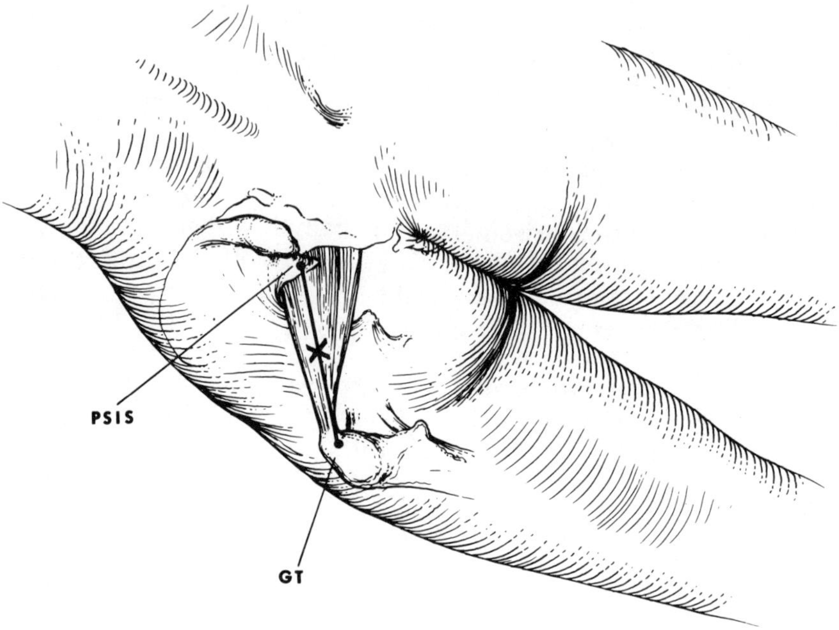

Innervation

Nerve to the Piriformis S1, S2.

Origin

From the front of the sacrum.

Insertion

Into the piriform fossa of the greater trochanter.

Position

The patient prone.

Electrode Insertion (X)

Insert the electrode deep to bone at the midpoint of a line between the posterior inferior iliac spine (SIS) and the posterior-superior margin of the greater trochanter (GT), then withdraw slightly.

Test Maneuver

Patient to externally rotate thigh.

Pitfalls

If the electrode is inserted too superficially it will be in the gluteus maximus; if inserted too caudally it will be on the obturator internus or gemelli.

Comment

(a) The sciatic nerve may be entrapped as it crosses this muscle (piriformis syndrome).
(b) Involved in lesions of S1 and S2. This muscle is innervated directly from these roots.

QUADRATUS FEMORIS

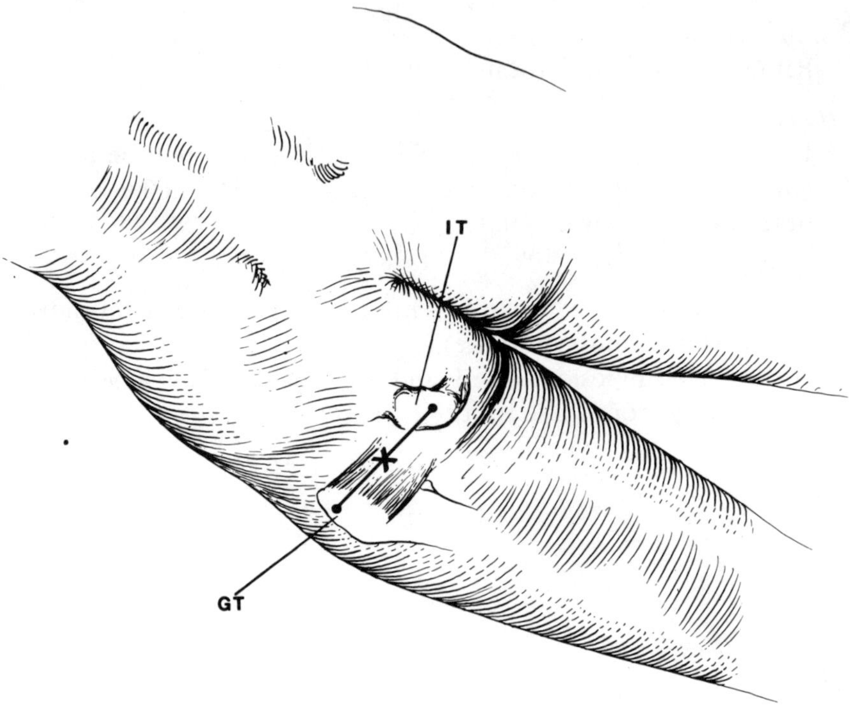

Innervation

Nerve to the Quadratus Femoris, Sciatic Nerve (Tibial Portion), L4, *L5, S1.*

Origin

From the ischial tuberosity deep to the hamstrings.

Insertion

Into the quadrate tubercle of the femur.

Position

The patient prone.

Electrode Insertion (X)

Midway between the greater trochanter (GT) and the ischial tuberosity (IT). The electrode should contact bone and then be withdrawn slightly.

Test Maneuver

Patient to externally rotate thigh.

Pitfalls

If the electrode is inserted too superficially it will be in the gluteus maximus or medius; if inserted too distally it will be in the hamstring group.

Comment

Involved in lesions of sciatic nerve at its junction with the sacral plexus or L4, L5, S1 roots.